Object Detection with Deep Learning Models

Object Detection with Deep Learning Models discusses recent advances in object detection and recognition using deep learning methods, which have achieved great success in the field of computer vision and image processing. It provides a systematic and methodical overview of the latest developments in deep learning theory and its applications to computer vision, illustrating them using key topics, including object detection, face analysis, 3D object recognition, and image retrieval.

The book offers a rich blend of theory and practice. It is suitable for students, researchers and practitioners interested in deep learning, computer vision and beyond and can also be used as a reference book. The comprehensive comparison of various deep learning applications helps readers with a basic understanding of machine learning and calculus grasp the theories and inspires applications in other computer vision tasks.

Features:

- A structured overview of deep learning in object detection
- A diversified collection of applications of object detection using deep neural networks
- An emphasis on agriculture and remote-sensing domains
- Exclusive discussion on moving object detection

Object Detection with Deep Learning Models

Principles and Applications

Edited by
S. Poonkuntran
Rajesh Kumar Dhanraj
Balamurugan Balusamy

CRC Press
Taylor & Francis Group
Boca Raton London New York

CRC Press is an imprint of the
Taylor & Francis Group, an **informa** business

A CHAPMAN & HALL BOOK

First edition published 2023
by CRC Press
6000 Broken Sound Parkway NW, Suite 300, Boca Raton, FL 33487-2742

and by CRC Press
4 Park Square, Milton Park, Abingdon, Oxon, OX14 4RN

CRC Press is an imprint of Taylor & Francis Group, LLC

Library of Congress Cataloging-in-Publication Data
Names: Poonkuntran, S., editor. | Dhanraj, Rajesh Kumar, editor. |
  Balusamy, Balamurugan, editor.
Title: Object detection with deep learning models : principles and
  applications / edited by S Poonkuntran, Rajesh Kumar Dhanraj,
  Balamurugan Balusamy.
Description: First edition. | Boca Raton : Chapman & Hall/CRC Press, 2023.
  | Includes bibliographical references and index.
Identifiers: LCCN 2022015567 (print) | LCCN 2022015568 (ebook) | ISBN
  9781032074009 (hardback) | ISBN 9781032349244 (paperback) | ISBN
  9781003206736 (ebook)
Subjects: LCSH: Computer vision. | Pattern recognition systems. | Deep
  learning (Machine learning)
Classification: LCC TA1634 .O255 2023 (print) | LCC TA1634 (ebook) | DDC
  006.3/7--dc23/eng/20220725
LC record available at https://lccn.loc.gov/2022015567
LC ebook record available at https://lccn.loc.gov/2022015568

ISBN: 978-1-032-07400-9 (hbk)
ISBN: 978-1-032-34924-4 (pbk)
ISBN: 978-1-003-20673-6 (ebk)

DOI: 10.1201/9781003206736

Typeset in Palatino
by SPi Technologies India Pvt Ltd (Straive)

Contents

Editors

Poonkuntran Shanmugam earned a BE degree in Information Technology from Bharathidasan University, Tiruchirapalli, India; and MTech and PhD degrees in Computer and Information Technology from Manonmaniam Sundaranar University, Tirunelveli, India. He is presently with VIT Bhopal University, Madhya Pradesh, India as Professor & Dean for the School of Computing Science and Engineering. He has more than a decade of experience in teaching and research and successfully executed three funded research grant projects from the Indian Space Research Organization, Defense Research Development Organization, and Ministry of New and Renewable Energy, Government of India, to the tune of 1.10 Crores. He received two seminar grants from Anna University, Chennai, and the All India Council for Technical Education-Indian Society for Technical Education for the tune of 4 Lacs.

He has published more than 80 technical publications, authored 6 books and 2 chapters. He is the recipient of Cognizant Best Faculty Award 2017–18 and served as a State Level Student Coordinator for Region VII, CSI, India in 2016–17. He is a lifetime member of IACSIT, Singapore, CSI, India, and ISTE, India. His research areas of interests include information security, computer vision, artificial intelligence, and machine learning.

Dr Rajesh Kumar Dhanraj is a Professor in the School of Computing Science and Engineering at Galgotias University, Greater Noida, India. He earned a BE degree in Computer Science and Engineering from the Anna University Chennai, India in 2007, then an MTech from the Anna University Coimbatore, India in 2010 and a PhD in Computer Science from Anna University, Chennai, India, in 2017. He has contributed to 30+ authored and edited books on various technologies, 21 Patents and 53 articles and papers in various refereed journals and international conferences and contributed chapters to books. His research interests include Machine Learning, Cyber-Physical Systems and Wireless Sensor Networks. He is a senior member of the Institute of Electrical and Electronics Engineers (IEEE), member of the Computer Science Teacher Association (CSTA), and the International Association of Engineers (IAENG). He is an associate editor and guest editor for reputed journals. He is an Expert Advisory Panel Member of Texas Instruments Inc., USA.

Balamurugan Balusamy is currently an Associate Dean Student in Shiv Nadar University, Delhi-NCR. Prior to this assignment he was Professor, School of Computing Sciences & Engineering and Director of International Relations at Galgotias University, Greater Noida, India. His contributions focus on Engineering Education, Block Chain and Data Sciences. His Academic degrees and twelve years of experience working as a faculty member in a global University like VIT University, Vellore, has made him more receptive and prominent in his domain. He has 200 plus high impact factor papers in Springer, Elsevier and IEEE. He has done more than 80 edited and authored books and collaborated with eminent professors across the world from top QS ranked universities.

Prof. Balamurugan Balusamy has served up to the position of associate professor in his 12 years stint with VIT University, Vellore. He completed his Bachelors, Masters and PhD degrees at top premier institutions in India. His passion is teaching, and he adapts different design thinking principles while delivering his lectures. He has published 80+ books about various technologies and visited over 15 countries for his technical courses. He has several top-notch conferences in his resume and has published over 200 quality journal articles, conferences and book chapters combined. He serves in the advisory committee for several start-ups and forums and does consultancy work for industry on Industrial IOT. He has given over 195 talks at various events and symposiums.

List of Contributors

S. Aanjanadevi
Alagappa University
Tamil Nadu, India

S. Aanjankumar
School of Computing Science and
Engineering, VIT Bhopal University,
India

J.V. Anchitaalagammai
Velammal College of Engineering &
Technology
Madurai, India

M. Arun Anoop
Royal College of Engineering and
Technology
Akkikkavu, Thrissur, Kerala

M. Brindha
NIT Tiruchirappalli
Tamil Nadu, India

Sachi Choudhary
University of Petroleum & Energy Studies
Dehradun, India

R. Deepalakshmi
Velammal College of Engineering and
Technology, Viraganoor
Tamil Nadu, India

K. Ganesh Babu
Chendhuran College of Engineering &
Technology
Pudukottai, India

T. Gladima Nisia
AAA College of Engineering and
Technology
Sivakasi, India

P.R. Hemalatha
Velammal College of Engineering &
Technology
Madurai, India

P. Karthikeyan
Velammal College of Engineering and
Technology
Madurai, Tamil Nadu, India

Amit Kumar
IIIT Kota
Rajasthan, India

M. Manikandakumar
Thiagarajar College of Engineering
Tamil Nadu, India

N. Mohammed Raffic
Nehru Institute of Technology
Coimbatore, India

C. Nandhini
NIT Tiruchirappalli
Tamil Nadu, India

V. Palanisamy
Alagappa University
Tamil Nadu, India

S. Poonkuntran
School of Computing Science and
Engineering, VIT Bhopal University,
Madhya Pradesh, India

D. Prabha
Sri Krishna College of Engineering and
Technology
Coimbatore, India

S. Rajesh
Mepco Schlenk Engineering College
Sivakasi, India

E. Ramanujam
National Institute of Technology Silchar,
    Assam, India

P. Ravikumaran
Fatima Michael College of Engineering &
    Technology
Madurai, Tamil Nadu, India

A.S. Renugadevi
Kongu Engineering College
Tamil Nadu, India

K. Sangeetha
Panimalar Engineering College
Chennai, India

T. Shanmuga Priya
Vuram Technologies
India

Gargeya Sharma
University of Petroleum & Energy Studies
Dehradun, India

Rashmi Sharma
University of Petroleum & Energy Studies
Dehradun, India

Vivek Sharma
MNIT Jaipur
Rajasthan, India

Ankit Shrivastava
School of Computing Science and
    Engineering
VIT Bhopal University, India

S. Srinivasan
Nehru Institute of Technology
Coimbatore, India

T. Suba Nachiar
Velammal College of Engineering &
    Technology Madurai, India

S. Thirumurugaveerakumar
Panimalar Engineering College Chennai,
    India

K. Valarmathi
P.S.R Engineering College
Sivakasi, India

R. Vijayalakshmi
Velammal College of Engineering and
    Technology
Tamil Nadu, India

K. Vimala Devi
Vellore Institute of Technology
Vellore, India

Monika Vyas
IIIT Kota
Rajasthan, India

1

Introduction: Deep Learning and Computer Vision

A.S. Renugadevi

Kongu Engineering College, Tamil Nadu, India

CONTENTS

DOI: 10.1201/9781003206736-1

1.1 Introduction to Deep Learning

1.1.1 Deep Learning

Artificial intelligence enables computers to mimic human behavior. The machine learning and deep learning concepts are covered under artificial intelligence. In machine learning, the algorithm for feature extraction should be given by humans, whereas in deep learning, the feature extraction is done automatically by the perception of neurons. The emergence of deep learning is illustrated in Figure 1.1.

The classification, clustering and predicting of the result by the neural networks are done in deep learning. The central part of deep learning is neural networks. The patterns are identified by using neural networks. The neural networks can be designed as a set of algorithms for predicting results. Nothing has emerged in deep learning newly; instead, due to the exponential increase in processing capacity, both machine learning and deep learning came into existence.

The neurons in the brain of humans are data carriers, and billions of neurons are connected with one another. Based on the logic of neurons in the human brain, the neurons are designed in the system. The artificial neural network is created with the help of neurons. Some neurons are going to act as input collectors, some neurons are going to act as output displayers, and some neurons are used in the processing of input.

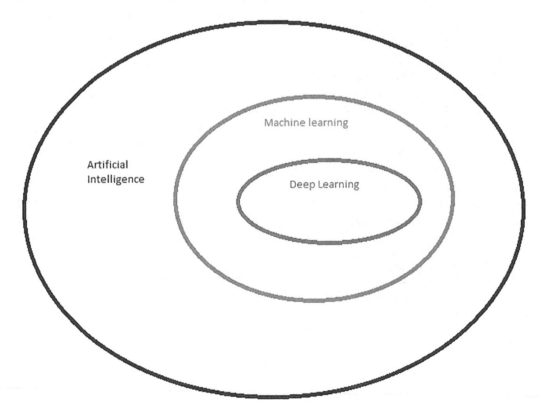

FIGURE 1.1
Emergence of deep learning.

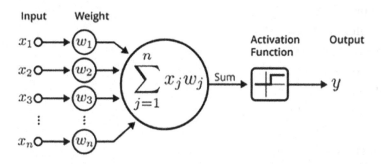

FIGURE 1.2
Illustration of an artificial neuron.

In the artificial neural network, the neuron plays a major role. The structure of an artificial neuron consists of inputs from x_0 through x_n and weights w_1 through w_n. Each input value is passed to the summation function. After that, the summed value obtained is passed to the activation function, and output y is generated. The structure of the neuron is given in Figure 1.2.

1.1.2 Machine Learning and Deep Learning

Machine learning can be called an approach for achieving artificial intelligence. It is the method of using algorithms to trace the data, learn from it, and make predictions over some things in the real world. Deep learning can be taken as a technique for implementing machine learning. The difference between machine learning and deep learning is explained in Table 1.1.

1.1.3 Types of Networks in Deep Learning

The neural networks that come under the category of machine learning help to predict the patterns by learning from the data and utilizing the neurons effectively. The types of neural networks are classified according to various features:

TABLE 1.1

Differences between Machine Learning and Deep Learning

Machine Learning	Deep Learning
Small amount of data is needed to provide accuracy.	Large amount of data is needed for training.
It requires low system specifications.	It requires high system specifications.
The given problem is divided into multiple tasks, and each task is solved independently. Finally, the results are combined.	The given problem is solved fully as a node-to-node problem.
The time needed for training the model is low.	The time needed for training the model is high.
But for testing the data with the model, the time required is high.	Here, less time is needed to test the data with the model.

Types of connection

    Static feedforward networks

    Dynamic feedback networks

Topology of networks

    Single-layer neural networks

    Multilayer neural networks

    Recurrent neural networks

Learning methods

    Supervised Learning

    Unsupervised Learning

    Reinforcement Learning

1.1.3.1 Connection Type of Networks

1.1.3.1.1 Static Feedforward Networks

Feedforward neural networks can be named deep feedforward networks or multilayer networks. This neural network is the most perfect deep learning model used. Feedforward networks work on the principle that the function s is approximated as s. For classifying the data or images with that of the networks, $t = s * (u)$ connects the input u to the output t. Also, a mapping $t = f(u; \theta)$ tells that the approximation happens with the value of θ. In feedforward networks, the feedback connection is not considered; only the inputs are considered for the output evaluation. The data flow from the input u is passed to the various calculations under the function f, and the output t is obtained. Figure 1.3 shows the static feedforward networks.

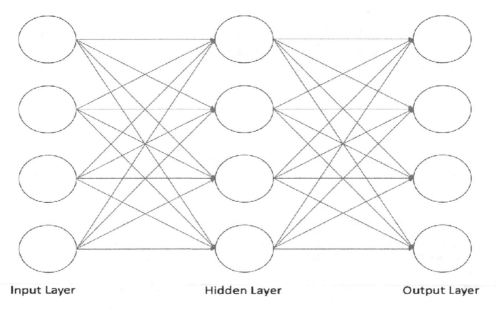

Input Layer   **Hidden Layer**   **Output Layer**

FIGURE 1.3
Feedforward networks.

Applications:

- Classification
- Speech recognition
- Face recognition
- Computer vision

1.1.3.1.2 *Dynamic Feedback Neural Networks*

In dynamic neural networks, the data can flow in two directions, namely the forward and backward directions. The most enhanced neural networks are the dynamic networks, and if any error occurs in the flow of data, it is tedious to get corrected. The various changes will take place in the states of the network until the equilibrium point is reached in the network. The changes in the input lead to the variation in the equilibrium point, but the network has to be at the same point till the variations occur in the equilibrium point. This type of network can be called a recurrent or interactive network. The feedback mechanism which is used in the networks is helpful in the process of addressing the content memories. The dynamic neural networks are shown in Figure 1.4.

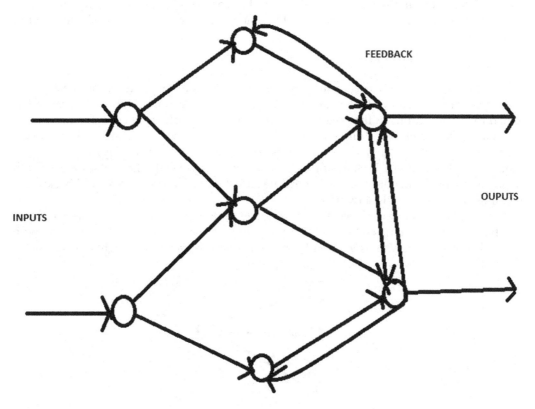

FIGURE 1.4
Feedback neural networks.

Applications of feedback neural networks:

- Word Processing
- Speech recognition
- Tagging an image
- Process of detecting sentiments
- Translation

1.1.3.2 Topology-based Neural Networks

1.1.3.2.1 Single-layer Neural Networks

The perceptron is one of the first samples of a single-layer neural network. The perceptron would return a function supported inputs, again, supported single neurons in the physiology of the human brain. The logic gates satisfying the individual functionality can be called a model of a perceptron in some cases. Based on the weighted inputs, the perceptron may send data or not. The type in which the single-layer network works out is the single-layer binary linear classifier, which helps to separate the input data as one of the two types.

Feedforward networks include single-layer neural networks because data flows only in one direction. That is, in single-layer networks, data comes from the input layer to the output layer; it does not consider the feedback from the output layer. Also, a single-layer network is different from the network that uses the backpropagation and the gradient descent along with the functions. The structure of the single-layer neural network is depicted in Figure 1.5.

1.1.3.2.2 Multilayer Neural Networks

The multilayer neural network is the one in which information enters the network and is sent through various layers of neurons in the network. Every node present in layer 1 is joined to other neurons in the next layer, and similarly, layer 2 is joined to the neurons in the consequent layer. So a fully connected network is formed. There are multiple layers hidden between the input and output layers. There are more than two layers between the input and output layers. Unlike single-layer neural networks, the flow of data is in both directions. That means the forwarding of data as in feedforward networks and backwarding of data as in feedback networks.

The inputs given to the network are multiplied with weights and sent to the activation function. The loss is reduced by modifying the weights and activation function along with the backpropagation. Values learned by the machines are taken as weights in the neural

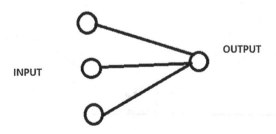

FIGURE 1.5
Single-layer neural network.

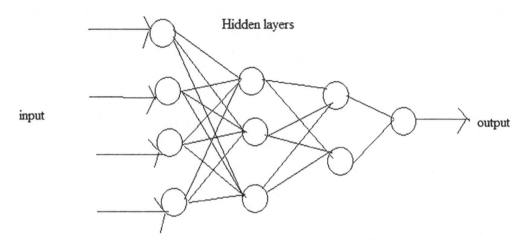

FIGURE 1.6
Multilayer neural network.

networks. The network gets self-adjusted based on the variation with the outputs predicted and the inputs trained in the network. The activation functions used are nonlinear and they are then sent to the softmax function. Figure 1.6 depicts the multilayer neural network.

Applications of multilayer perceptron

- Machine translation
- Recognition of speech
- Classification of complex images

1.1.3.2.3 Recurrent Neural Networks

The design motive of the recurrent network is to predict the output based on the input that is given as feedback to the network. The first layer in the network is the simple forwarding layer, and after that it is followed by the recurrent neural network in which the data or information residing in the memory is used. Forward propagation can be applied in layer 1, and in layer 2, the information is stored in the memory for use in the future. The incorrect prediction may be corrected by making changes with the help of the learning rate. That will be helpful in the correct prediction of the data as well as the images at the time of backpropagation. The recurrent neural network is shown in Figure 1.7:

Applications of recurrent neural networks

- Word Processing
- Speech recognition
- Tagging an image
- Process of detecting sentiments
- Translation

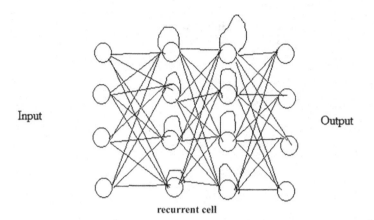

Input   Output

recurrent cell

FIGURE 1.7
Recurrent neural network.

1.1.3.3 Learning Methods

1.1.3.3.1 Supervised Learning

The most common form of deep learning is supervised learning. The set of images or data can be taken as a training set, and it is given as input to the network with the aspect of training the network. For every input, there will be a labeled corresponding output, such that the input can be processed and the desired output reached. As an example, the images are classified into X different classes. So that it needs a training set of images and a validation set of images. The training set can be written as {(r1,s1), (r2,s2),…..(rx,sx)}, where the input is ri and output is si [1]. Then the images can be trained by using the minimization of a cost function that will connect the output along with the correct input. The trained images are given to the model, and the model predicts the output. Figure 1.8 shows the method of supervised learning.

1.1.3.3.2 Unsupervised Learning

Unlike supervised learning, in unsupervised learning the training data or image set is not labeled for finding the classes or classifying the classes. So the network model finds the common characteristics among the data or images and clubs the data based on the knowledge of the model. The method of unsupervised learning is illustrated in Figure 1.9:

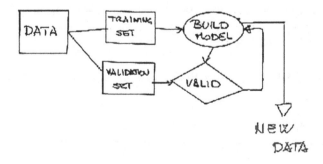

FIGURE 1.8
Supervised learning.

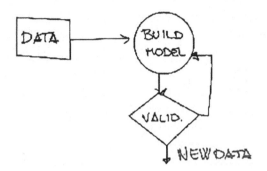

FIGURE 1.9
Unsupervised learning.

1.1.3.3.3 Reinforcement Learning

In reinforcement learning, without the training dataset, the suitable decision is taken on its own with the help of its experience. That decision will help to receive the reward in certain situations. It is achieved by using the different types of machines or software, whatever they may be, but the solution is only to reach the best path or behavior. How reinforcement learning varies from supervised and unsupervised learning is that the training data along with the correct solution are available in the two types of learning, whereas the training data are not available in reinforcement learning. So the reinforcement agent has to decide what to do to perform the allocated work [2]. The diagram in Figure 1.10 gives the idea of reinforcement learning.

1.2 Convolutional Neural Networks

1.2.1 Description of Five Layers of General CNN Architecture

The type of feedforward artificial neural network is the convolutional neural network. It can also be called ConvNet [3, 4, 5]. The multiple layers of artificial neurons are present in

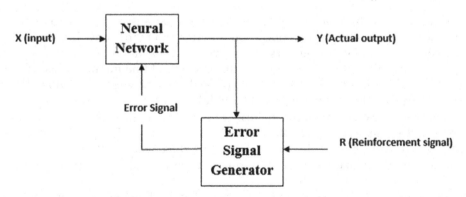

FIGURE 1.10
Reinforcement learning.

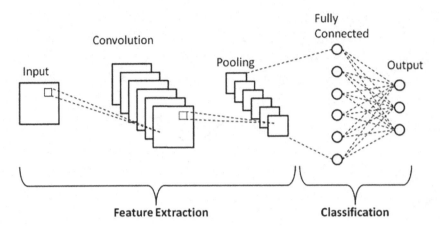

FIGURE 1.11
Convolutional neural networks.

the convolutional neural networks. The visual images are analyzed by this convolutional neural network. The neural networks used to analyze the visual images are also called shift invariant or space invariant artificial neural networks (SIANN) that scan the layers of convolutional neural networks and translation invariance characteristics on the basis of shared weight architecture. The translation invariance characteristics can be named feature maps. The convolutional neural network consists of convolutional layers (one or more) along with the pooling layer and the fully connected layer (one or more). The architecture of the convolutional neural networks is shown in the Figure 1.11.

CNN is a specific version of the neural network designed to operate with one-dimensional, two-dimensional, and three-dimensional data and images [6].

1.2.1.1 Input Layer

The whole CNN input depends on the input layer. The images are represented as the pixel matrix in the neural network.

1.2.1.2 Convolutional Layer

The name of the convolutional neural networks is given because of the convolutional layers in the network. The convolution operation is performed in the convolutional layer.

In the convolutional neural network, the convolution operation can be done by multiplying the input with that of the set of weights as in the old neural networks. When the two-dimensional input is taken, the two-dimensional array of weights called kernel or filter is multiplied with the two-dimensional input [6].

When the kernel used is smaller than the input data, the dot product can be said to multiply the small kernel-size input patch with the small kernel. The single value can be obtained by adding the results obtained in the dot product, which is the elementary multiplication of the kernel-size patch of the input and the kernel. Since the single value is obtained, it is called a scalar product.

The filter size should be smaller than the original input, then only the same size of the filter can be repeatedly multiplied by the input array at multiple points in the input. The

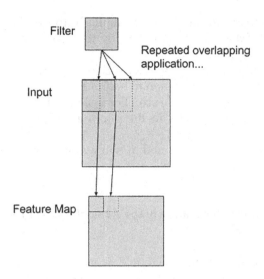

FIGURE 1.12
Extraction of feature map.

filters can be applied to each small size of the input either in the direction of top to bottom or in the direction of left to right.

The repeated application of the filter to the small size of the image is a very useful technique for identifying an exact feature in the input images. If the filter is applied in a similar manner to the entire image, then the features can be easily identified throughout the image. This concept is called translation invariance.

The single value is obtained as a result of multiplying the filter value with the small patch of the input. But if the filter value is applied all over the array of inputs, then the two-dimensional array of values is obtained. Those values seemed to be a filtering in input values. The output obtained by multiplying the filter with the input array is known as a feature map. After getting the feature map, the feature map is applied to the nonlinearity function ReLU. The feature map extraction is shown in Figure 1.12.

The convolution operation is actually called a cross-correlation operation in technical terms. The kernel value is rotated before applying to the input sometimes. The cross correlation in deep learning is known as convolution operation.

1.2.1.3 Pooling Layer

In the convolutional neural networks, after the convolutional layer, the pooling layer is added. The output from the convolutional layers is passed to the ReLU function, which will apply the nonlinearity to the output of the convolutional layer (i.e., the feature maps). So the ReLU function is added in between the convolutional layer and the pooling layer [6, 7].

The use of the pooling layer may be repeated after each convolutional layer in the neural network. Usage of the pooling layer may be decided based on the application. The pooling layer is applied to the feature maps of the convolutional layer, so the pooled feature maps are created in the same number from the pooling layer.

The pooling layer will perform a pooling operation, according to how the filter is going to apply to the feature maps. Normally, the filter size is comparably lesser than that of the value of input in order to create feature maps. Similarly, the pooling operation size is also small compared to that of the feature maps. Exactly, the pooling operation size is 2*2 pixels which is applied to the 2 pixels stride.

The pooling layer will use the 2 factor as a size of features extracted in the map. The reduction is carried out in each dimension to half of the original size, and as a result, the pixel value is reduced to 1/4 of the total size. For instance, if the total number of pixels is 36 (6*6 matrix), the number of pixels in the pooling layer is reduced to 9 pixels (3*3 matrix).

The pooling operation can be performed in two ways: Average pooling and maximum pooling.

Average pooling:
Each patch's average value of the feature maps is calculated [6]. The average pooling function is shown in Figure 1.13.

Maximum pooling (or max pooling):
Each patch's maximum value of the feature maps is calculated [6]. The max pooling function is shown in Figure 1.14.

The finalized version of the features identified in the input is computed in the pooled feature maps. The downsampled feature maps are the result of the usage of the pooling layer. The translation invariance is calculated from the convolutional layer, which is converted to local translation by means of the pooling layer.

The amount of the transfer taken place in the pooling layer is very small, so the output from the pooled layer is mostly not changeable. The approximations obtained are from invariant to small transactions.

1.2.1.4 Fully Connected Layers

It takes the input from the final pooling layer. The output from the final pooling layer is flattened. The flattening of inputs means that the output in the three-dimensional matrix is unrolled into individual vectors [6, 8]. The flattening concept is given in Figure 1.15.

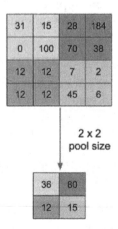

FIGURE 1.13
Average pooling function.

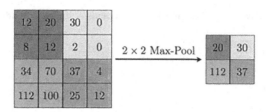

FIGURE 1.14
Max pooling function.

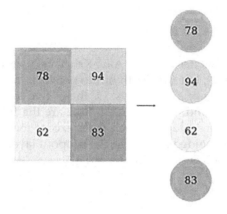

FIGURE 1.15
Flattening.

The flattened value obtained in the above figure is given as input to the fully connected layer. The result of the fully connected layer is sent to the final layer, which uses the softmax activation function for classifying the results. The results can be classified into various classes.

1.2.1.5 Output Layer

The output is then generated through the output layer generates the output and the error checking is also performed. As a result, the loss function is computed and also gradient error is calculated.

1.2.2 Types of Architecture in CNN [9]

1.2.2.1 LeNet-5

The LeNet-5 architecture was designed by LeCun et al. in 1998. It is the earliest model used for classifying handwritten numbers and digits. The LeNet-5 architecture is thus named because it uses five layers. Three of the five layers are convolutional layers, along with pooling layers for each convolutional layer. The two layers are the fully connected layers. Finally, the softmax classifier is used to classify the images into respective classes. This architecture is most popular because this is a very straightforward approach [10, 11]. Figure 1.16 shows the architecture of LeNet-5.

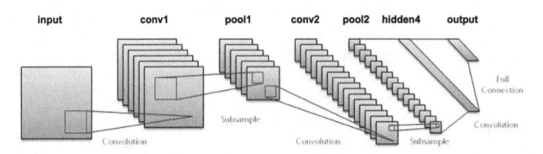

FIGURE 1.16
LeNet-5.

1.2.2.2 AlexNet

The AlexNet architecture was designed by Alex Krizhevsky et al. in 2012. AlexNet has a similar architecture to LeNet, but the depth of the network in AlexNet is increased. The AlexNet architecture consists of eight layers. Of these, five are the convolutional layers with the max pooling layer, and the remaining three are the fully connected layers. The ReLU activation functions are added in each layer except the output layer. The overfitting in the network can be avoided by adding the dropout layers in the network [10]. The AlexNet architecture diagram is shown in Figure 1.17.

1.2.2.3 ZFNet

ZFNet was designed in 2013 in order to optimize the performance of AlexNet. The depth of the networks can be increased by adding the extra filters in the same structure as AlexNet. Instead of increasing the filter size, the number of filters or kernels is increased to optimize the performance [10]. The architecture diagram is given in Figure 1.18.

1.2.2.4 GoogLeNet/Inception

The architecture of GoogLeNet differs from the other architectures in the way that it uses the 1*1 convolution and global average pooling to create the deeper networks. The number of parameters used in the convolution is decreased so that the deepness of networks gets

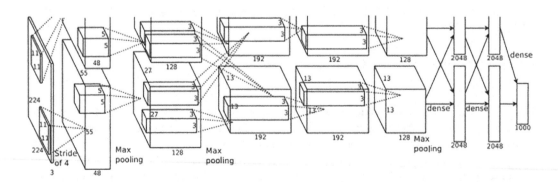

FIGURE 1.17
AlexNet.

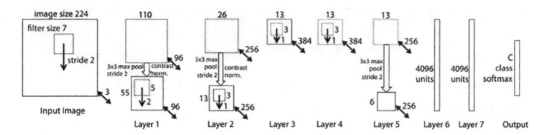

FIGURE 1.18
ZFNet.

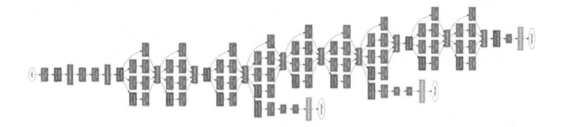

Convolution
Pooling
Softmax
Other

FIGURE 1.19
GoogLeNet.

increased. The accuracy of the classification is increased by means of the global average pooling. The fully connected layer with the ReLU activation function is used, and also the dropout layer is used for regularization [10]. The softmax classifier is used for the classification of images or data. Figure 1.19 shows the block diagram of GoogLeNet.

1.2.2.5 VGGNet

VGGNet was designed by Simonyan and Zisserman in 2014. VGGNet architecture has a total of 16 convolutional layers. The number of filters is increased as in AlexNet. The 3*3 filters are added to increase the depth of the network. The three fully connected layers are added at the end after the pooling layers [10, 12]. The VGGNet architecture is depicted in Figure 1.20.

1.2.2.6 ResNet

The ResNet was designed by Kaiming He et al. in 2015. ResNet is introduced to get rid of the vanishing gradient. The skip connection technique is used in the ResNet network. The skip connection works in the way that particular training is skipped from a few layers, and the remaining is connected to the output layer [10]. The architecture of ResNet is shown in Figure 1.21.

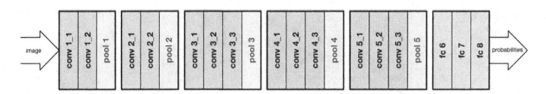

FIGURE 1.20
VGGnet.

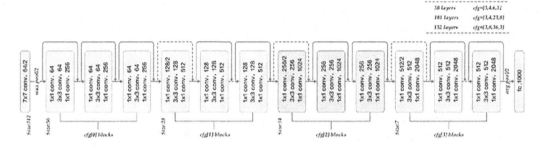

FIGURE 1.21
ResNet.

1.2.3 Applications of Deep Learning

1. Automatic text generation – The learning of text is done and the new text is also framed with the help of the model. The model helps to learn how to punctuate, spell and frame new sentences and also the style is captured sometimes.

2. Healthcare – Various diseases can be diagnosed and also treated earlier.

3. Automatic machine translation – The translation of text in one language is converted into another language automatically. The text may be words or sentences.

4. Image recognition – Objects and people are recognized and identified with the help of deep learning.

5. Predicting earthquakes – Deep learning trains the model to predict earthquakes earlier.

6. Industrial applications – Object detection and localization, sorting, robotics, quality control and inspection, packaging.

7. Retail applications – analytics, warehouse management, theft prevention, intelligent barcode scanners, monitoring and distribution control.

8. Entertainment/gaming – gesture recognition, user identification, emotional feedback, experience monitoring, advance analytics.

9. Smart homes – vacuum cleaners, automatic lawn movers, intrusion and hazard detection, smart lights, ovens, refrigerators.

10. Agriculture – weed control, fruit harvesting, autonomous tractors and combines.

11. Smart cities and infrastructure – parking, traffic monitoring, security monitoring, road inspection.

12. Food industry – sorting, quality control.

1.3 Image Classification, Object Detection and Face Recognition

The field of computer science, which deals with the creation of digital systems for processing, analyzing and visualizing the data in the way humans do, is called computer vision. Computer vision is training computers to understand images and to process them. Finally, with the training made, the computers are ready to retrieve the visual data and send the final results with the help of software algorithms. Computer vision aims to classify images, detect objects, and recognize faces. The tasks relevant to computer vision using deep learning are:

- Dataset creation
- Preprocessing
- Image classification
- Object detection
- Face recognition

1.3.1 Dataset Creation

A dataset is a collection of data and its related values. The dataset has both the parameters as time and subject. The dataset creation is a challenging task in deep learning. The data collection is a static process. The collection of data is over a period of time; labeling the data, training the model and results are found in deep learning. There are different types of datasets such as text data, image data, signal data, sound data, physical data, anomaly data, biological data, multivariate data, question-answering data and other data repositories.

The performance of deep learning is improved by improving the data. That means the addition of more data to train the model will be helpful in classifying the data.

Data acquisition is the process by which datasets are found for training the models. The two methods of data acquisition are:

1. Data generation
2. Data augmentation

Data generation is accomplished by crowdsourcing (connecting with people to collect the data) and synthetic data generation (computer-generated data).

Dataset creation also involves searching for open-source datasets, available on the internet. Many open source datasets, such as Kaggle, PlantVillage dataset, etc., are available. Also, the dataset can be collected directly from the companies, hospitals, etc. using the data gathering mechanism.

Data augmentation techniques increase the number of images in the training dataset by applying various operations such as flipping, rotation, etc.

Some of the image repositories are:

- Scikit-Image
- OpenCV
- Python Image Library (Pillow/PIL)

- Scipy
- SimpleITK
- Matplotlib
- Numpy
- Mahotas

1.3.2 Data Preprocessing

The preprocessing of the dataset involves both the text dataset and image dataset preprocessing. The text dataset preprocessing consists of the steps such as

- Removal of punctuation
- Lower casing
- Spelling correction
- Removal of frequent words
- Chat words conversion
- Removal of URLs
- Lemmatization
- Removal of rare words
- Stemming
- Removal of emoticons
- Conversion of emoji to words
- Removal of stopwords
- Removal of emoji
- Conversion of emoticons to words
- Removal of HTML tags

The preprocessing of image datasets consists of image resizing, noise removal, segmentation, and edge smoothing.

Image resizing is varying the size of the image. Unwanted noise can be removed from the images by using noise removal techniques. The particular part of the images can be segmented using segmentation. The edges of the images can also be smoothed using edge smoothing techniques.

1.3.3 Image Classification

The features extracted from the images for observing patterns in the dataset are helpful in image classification [13]. If an artificial neural network is used for image classification, then the classification process is very costly [14]. So CNN is used for the classification. There are different types of classification problems, such as single label classification and multilabel classification in supervised learning, unsupervised classification, video classification, and 3D classification.

The steps carried out in the classification process are as follows:

Step 1: Specific dataset should be chosen. Choose a dataset already available or create your own dataset.

Step 2: Import the necessary libraries needed for the classification.

Step 3: Prepare the training dataset by assigning the path and also create the categories. Also resize the images.

Step 4: Create the data in the training data set and shuffle the dataset. Assign the labels as well as features to the entire image.

Step 5: Normalize the X values and convert labels into categorical data. Split the X values and Y values for using it in CNN.

Step 6: Define the model, compile it and train the CNN model.

Step 7: Find the accuracy of the model in classifying the objects.

Examples of classification problems:

- CNN model to perform classification of dogs and cats photographs
- CNN model to perform labeling of photographs of the Amazon rainforest

Binary classification

- Identification of cancer in X-ray images

Multiclass classification

- Handwritten text can be classified using CNN
- The photograph of a face can be assigned a name by using CNN

1.3.4 Object Detection

Object detection may be referred to as object recognition; since it combines the two functionalities such as drawing a bounding box around each and every object, which needs to be identified in the images and then assigning a label to the identified object [13]. Image classification is a straightforward technique, whereas object detection also involves the localization of the objects.

For addressing object localization, region-based convolutional neural networks are used. R-CNNs are designed specifically for recognizing objects.

The YOLO model (you only look once) is also designed specifically for detecting objects in the images considering the speed and the real-time usage. The variation between the three tasks can be explained as follows:

Image classification: The type or class of the object can be identified in an image [15].

- *Input*: Single image or photograph is given as input.
- *Output*: A label of the class (corresponding to images)

Object localization: The presence of objects is located in an image and also a bounding box for indicating their exact location.

- *Input*: One or more objects can be present in the image or photograph.
- *Output*: One or more bounding boxes corresponding to objects along with the width, height and point.

Object detection: The presence of objects is located in an image and also bounding box for indicating their exact location and also the labeled classes of the exact objects should be given as output [16, 17, 18].

- *Input*: One or more objects can be present in the image or photograph.
- *Output*: One or more bounding boxes corresponding to objects along with the width, height and point and also the class type or label of the identified object.

The steps carried out in the object detection process are as follows:

Step 1: Specific images should be taken as input.

Step 2: The images should be divided into various regions.

Step 3: Each region should be considered as a separate image to work with.

Step 4: Send all the regions considered as individual images to the model, and classify the images into different types of classes.

Step 5: After the classes are identified for each region, all the regions are again combined to identify the objects in the original image.

Examples for object detection:

- Each object in a street scene should be identified by a bounding box, and also object should be labeled.
- Each object in an indoor photograph should be identified by a bounding box, and also object should be labeled.
- Each object in a landscape should be identified by a bounding box, and also object should be labeled.
- Object detection models for locating and detecting the kangaroos in the photographs [19, 20].

1.3.5 Face Recognition

Face recognition is the task in computer vision in which human faces are identified in photographs. Humans easily perform face detection, but it is a challenging problem for computers to recognize human faces. Face recognition becomes a nontrivial problem for computers to solve [21].

In face detection, the faces of different humans in the photograph should be located. The coordinates of the faces in the images should be represented by using the bounding box. The dynamic nature of the human face should be considered irrespective of the angle or orientation. Also, other parameters such as hair color, clothing, light levels, accessories, age and makeup should be considered.

There are two methods used for the recognition of faces. They are:

- Methods based on features – Detecting the faces with the help of handcrafted filters
- Methods based on images – Extracting the faces using the holistic learning from the entire image

The steps involved in the process of face recognition are:

Step 1: Images containing multiple faces should be given as input.

Step 2: One or more faces in the images should be located and marked with the bounding box.

Step 3: The face should be normalized and consistent with the database's photometrics and geometry.

Step 4: The features should be extracted for the recognition of faces from the face.

Step 5: The exact matching of the face with one or more faces stored in the database should be performed.

The three models frequently used for face recognition are multi-task cascaded convolutional neural network (MTCNN), the VGGFace2 model, and the FaceNet model.

The MTCNN model is the most used model for detecting faces with expressions. It was developed in 2016. As the name implies, the three neural networks are connected in a cascade way, which helps detect faces and facial landmarks in the images.

Face identification and verification can be performed by using the VGGNet2 model. VGG stands for Visual Geometry Group. The embedding of faces can also be detected using this model.

The FaceNet model is mainly used for feature extraction from the human face. It is also used for face identification and verification purpose.

References

1. https://towardsdatascience.com/derivative-of-the-sigmoid-function536880cf918e
2. https://www.medcalc.org/manual/tanh_function.php
3. Jie Wang and Zihao Li, "Research on Face Recognition Based on CNN," IOP Conf. Series: Earth and Environmental Science 170 (2018), 032110. DOI:10.1088/1755-1315/170/3/032110.
4. Keiron O'Shea, Ryan Nash An, "Introduction to Convolutional Neural Networks," arXiv:1511.08458v2 (2015).
5. https://towardsdatascience.com/a-comprehensive-guide-to-convolutional-neural-networks-the-eli5-way-3bd2b1164a53
6. Athanasios Voulodimos, Nikolaos Doulamis, Anastasios Doulamis, and Eftychios Protopapadakis, *"Deep Learning for Computer Vision: A Brief Review,"* Recent Developments in Deep Learning for Engineering Applications (2018). DOI:10.1155/2018/7068349.
7. https://learnopencv.com/image-classification-using-convolutional-neural-networks-in-keras/
8. https://www.tinymind.com/learn/terms/relu

9. https://medium.com/geekculture/a-2021-guide-to-improving-cnns-network-architectures-historical-network-architectures-d23f32afb1bd
10. A Ghosh, A Sufian, and F Sultana, "Fundamental Concepts of Convolutional Neural Network," *Recent Trends and Advances in Artificial Intelligence and Internet of Things* (2020). DOI:10.1007/978-3-030-32644-9_36.
11. Laith Alzubaidi, Jinglan Zhang, Amjad J. Humaidi, Ayad Al-Dujaili, Ye Duan, Omran Al-Shamma, J. Santamaría, Mohammed A. Fadhel, Muthana Al-Amidie and Laith Farhan, "Review of deep learning: concepts, CNN architectures, challenges, applications, future directions," *Journal of Big Data* (2021), DOI:10.1186/s40537-021-00444-8.
12. https://towardsdatascience.com/step-by-step-vgg16-implementation-in-keras-for-beginners-a833c686ae6c
13. Geert Litjens, Thijs Kooi, Babak Ehteshami Bejnordi, Arnaud Arindra Adiyoso Setio, Francesco Ciompi, Mohsen Ghafoorian, Jeroen A.W.M. van der Laak, Bram van Ginneken, Clara I. Sanchez, "A Survey on Deep Learning in Medical Image Analysis," arXiv:1702.05747v2 (2017).
14. https://en.wikibooks.org/wiki/Artificial_Neural_Networks/Activation_Functions
15. https://www.analyticsvidhya.com/blog/2021/01/image-classification-using-convolutional-neural-networks-a-step-by-step-guide/
16. Zhong-Qiu Zhao, Member, IEEE, Peng Zheng, Shou-tao Xu, and Xindong Wu, Fellow, IEEE, "Object Detection with Deep Learning: A Review," arXiv:1807.05511v2 (2019).
17. Zhixue Wang, Jianping Peng, Wenwei Song, Xiaorong Gao, Yu Zhang, Xiang Zhang, Longfei Xiao, and Li Ma, "Research Article A Convolutional Neural Network-Based Classification and Decision-Making Model for Visible Defect Identification of High Speed Train Images," *Journal of Sensors*, (2021), 5554920, DOI:10.1155/2021/5554920.
18. https://books.google.co.in/books?hl=en&lr=&id=10jpDwAAQBAJ&oi=fnd&pg=PP1&dq=deep+learning+and+computer+vision&ots=wHn2HtMBT2&sig=lNP7CXdDIy2Tk1BrcsTv6QJwXmM#v=onepage&q=deep%20learning%20and%20computer%20vision&f=false
19. Ajeet Ram Pathak, Manjusha Pandey, and Siddharth Rautaray, "Application of Deep Learning for object detection," *Procedia Computer Science* 132 (2018), 1706–1717, DOI: 10.1016/j.procs.2018.05.144.
20. https://www.upgrad.com/blog/ultimate-guide-to-object-detection-using-deep-learning/
21. KH Teoh, RC Ismail, SZM Naziri, R Hussin, MNM Isa and MSSM Basir, "Face Recognition and Identification using Deep Learning Approach," *Journal of Physics: Conference Series* 1755 (2021), 012006. DOI:10.1088/1742-6596/1755/1/012006.

2

Object Detection Frameworks and Services in Computer Vision

Sachi Choudhary, Rashmi Sharma, and Gargeya Sharma

University of Petroleum & Energy Studies, Dehradun, India

CONTENTS

DOI: 10.1201/9781003206736-2

2.1 Neural Networks (NNs) and Deep Neural Networks (DNNs)

One well-known machine learning methodology is deep learning. Learning the correlation between input and output data is done via neural network computational graphs. The concept of 'deep learning' evolves from the fact that NNs are highly flexible and can easily be arranged on top of each other to form deep computational graphs as per the need of the application.

2.1.1 Neural Networks

The stackable computational graphs used in deep learning are called neural networks, which refer to the term from neurobiology. But they hardly resemble the workings of our brains, not to be confused. They are a mathematical framework for learning representations from data. Any graph can be broken into smaller pieces which can be broken down until they reach their independent atomic component. In the case of neural networks, the smallest independent unit is called a neuron [1,2].

$$Wx + B = Z \tag{2.1}$$

$$f(Z) = Y \tag{2.2}$$

A single neuron is a group of two mathematical equations, as shown in Figure 2.1. The equation (2.1) is the most basic linear equation where x is our input and W and B are coefficients for the equation. This equation is responsible for learning the linear representations in the data. Learning to understand only linear relationships is not sufficient in most cases, as the real world contains so many irregularities, noise, and nonlinearity in the data. For learning nonlinear representations, equation (2.2) can be used in each neuron to wrap a

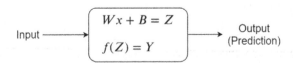

FIGURE 2.1
An artificial neuron.

function around the output from equation (2.1); these functions are called activation functions. Activation functions are discussed in more detail in later topics.

2.1.2 Single-Layer Perceptron (SLP)

The SLP was the first neural network model proposed by Frank Rosenblatt in 1958 [3]. It is among the earliest models to propose learning from data. The goal was to discover a linear decision function that classifies the output into binary classes (categories) with the use of the values in the weight vector (w) and the bias parameter (b) [4]. Figure 2.2 shows the single-layer perceptron. The inputs to the neural network are multiplied by their respective weight vector values, and their sum is further combined with a bias term to produce a single value output, equation (2.3). Such a single value output is passed as the input to the activation function assigned to the neuron, equation (2.4).

$$y = \sum x_i.w_i + b \tag{2.3}$$

$$y = \begin{cases} 1 \, if \, y \geq 0 \\ 0 \, if \, y < 0 \end{cases} \tag{2.4}$$

2.1.3 Multilayer Perceptron (MLP)

It can be seen that a perceptron is a linear function, so the trained single neuron will produce a straight line to classify the data. Will this work for complex nonlinear datasets? The answer is that many neurons are needed to optimally fit the training data. A multilayer perceptron has the same structure as a single-layer perceptron but contains two or more hidden layers. Hidden layers are collections of neurons that are not directly accessible by the input data; they act as intermediate processing units between the raw input and the final output. Typically, each neuron in the hidden layer is linked to every other neuron in adjacent layers, forming a denser connection between them and providing more

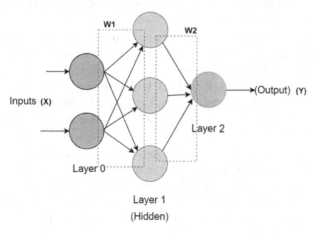

FIGURE 2.2
Single-layer perceptron (SLP).

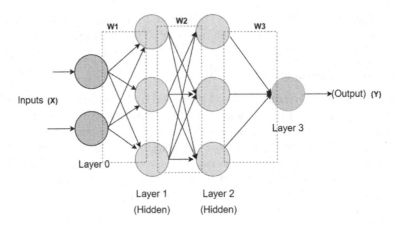

FIGURE 2.3
Multilayer perceptron (MLP).

computation to the expected output [4]. Figure 2.3 shows a multilayer perceptron with two hidden layers.

2.2 Activation Functions

Recalling the formation of a neuron, activation functions are applied to neurons in a layer during prediction. They convert the linear output into a nonlinear form. It is embedded after every perceptron, and it decides the activation of that neuron. There are several constraints to make activation work. Some of the primary constraints that turn a normal function into an activation function are [5]:

1. Continuity of function: They must be continuous and infinite in the domain. It must contain an output number for any input. There should be no restriction in its domain for it to fail to give an output value.

2. Monotonic in nature: They should never change direction. In other words, it is either always increasing or always decreasing. This constraint is not technically a requirement. Unlike functions with missing values, you can optimize non-monotonic functions. Unlike functions with missing values, one can optimize non-monotonic functions. Nevertheless, consider the implications of having multiple input values map to the same output value. It is not advisable to get such a result for learning.

3. Nonlinear in nature: One of the two equations in the neuron is sufficient to identify the linear representation in the data and build a linear prediction model. However, for nonlinear representations, there is no mathematical equation to counteract such nonlinear behavior in the data. Therefore, activation functions are kept nonlinear to promote learning of nonlinear correlations and their respective representations in the data.

The activation functions can be broadly divided into linear and nonlinear functions. Some of the most popular activation functions used in deep learning are.

2.2.1 Identity Function

In an identity function, also known as linear transfer function, the output is the same as the input, equation (2.5) (Figure 2.4).

$$f(x) = x \qquad (2.5)$$

The most used nonlinear activation functions are:

2.2.2 Sigmoid Function

Also known as logistic activation function, equation (2.6). The sigmoid is extremely popular with classification tasks because it smoothly eliminates infinite amounts of input into an output between 0 and 1 (Figure 2.5).

$$\sigma(z) = \frac{1}{1 + e^{-z}} \qquad (2.6)$$

2.2.3 Softmax Function

The conversion of input values into probability values is done by softmax function equation (2.7). It is often used at the output layer of a classification model where prediction of the class between more than two classes is required.

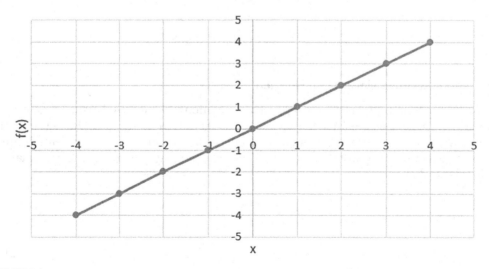

Linear activation function

FIGURE 2.4
Linear activation function.

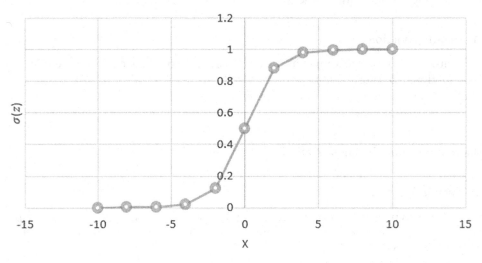

FIGURE 2.5
Sigmoid activation function.

$$\sigma(x_j) = \frac{e^{x_j}}{\sum_i e^{x_i}} \qquad (2.7)$$

2.2.4 Tanh Function

The tanh function is similar to the sigmoid (equation 2.8), except that it squishes the infinite range of input values from −1 to 1, as opposed to 0 to 1 by the sigmoid function (Figure 2.6).

$$\tanh(x) = \frac{\sinh(x)}{\cosh(x)} = \frac{e^x - e^{-x}}{e^x + e^{-x}} \qquad (2.8)$$

2.2.5 ReLU (Rectified Linear Unit) Function

The ReLU function is by far the most popular activation function used inside most neural networks because it results in a constant value as its output. It activates the neuron only if the output value is greater than zero, equation (2.9). The ReLU function is quite simple, in that it converts all negative numbers to 0 and no change to positive numbers, equation (2.10) (Figure 2.7).

$$z = \max(0, x) \qquad (2.9)$$

$$ReLU(x) = \begin{cases} 0 \ if \ x < 0 \\ x \ if \ x > 0 \end{cases} \qquad (2.10)$$

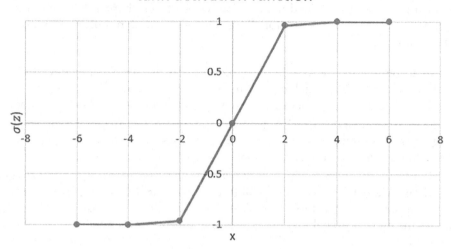

FIGURE 2.6
Tanh activation function.

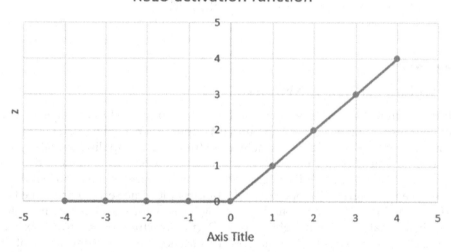

FIGURE 2.7
ReLU activation function.

2.3 Loss Functions

The loss function, also known as the error function, comes under the umbrella body of deep learning that encapsulates the idea of how well a model is performing in relation to how it should be. It is used to measure the incorrect predictions made by the neural network with respect to its true class. This is an optimization problem. Minimizing the loss by optimizing the parameters will yield better accuracy of the model. The results of various loss functions for the same prediction will be different, and have significant consequences on the performance of the trained model. The scope of this chapter does not allow for a full explanation of the various loss functions. However, some popular loss functions are explained below [6].

1. Mean absolute error (MAE): To work out how far the actual value deviates from that predicted by the model, this formula is used. The mean absolute error maintains the same scale of error as the values by adjusting the standard deviation.

2. Mean squared error (MSE): This calculates the square of the difference between the target price and the predicted value. This increases the scale of error by squaring the value and making the model more sensitive to higher loss values.

3. Cross-entropy: Generally, this is used in classification problems as it calculates the difference between two probability distributions. Classifying a single training example with respect to all available classes would mean that whichever class shows the highest probability of representing the example as belonging to the corresponding class. Ideally, the aim is to get a 100 percent accurate prediction for the correct class and 0 percent for the rest during training and learn to estimate this score.

There are many more loss functions than discussed that one can find and use, depending on the type of problem and their optimization approach.

2.4 Convolutional Neural Networks

An artificial neural network (ANN) or multilayer perceptron (MLP) is a layered arrangement of neurons having weight (w) and bias (b). Inputs are passed to each neuron, which is then multiplied by the weights and activation functions are applied to make the result of that layer nonlinear. A convolutional neural network (CNN) is an advanced version of a regular neural network or MLP designed to improve the processing of spatial data (also known as data with a grid-like topology) [6]. For example, time-series data can be thought of as 1D gridded data formed from aggregating those values at regular time intervals. Similarly, image data can be thought of as a 2D grid structure formed from pixel values and their 2-dimensional position on the grid. This section covers the development of convolutional neural networks (CNNs), which have produced better results for images and computer vision applications than MLPs.

2.4.1 CNN Architecture and its Components

Like a regular NN, the input to a CNN model is also an input image or a feature vector, which is transformed through a set of hidden layers and nonlinear activation functions. Each layer consists set of neurons that are connected with each neuron of the previous layer. The output layer is a fully connected layer and performs the classification. However, NN performs directly on the raw pixels and does not perform scaling on the image. In a CNN, layers are arranged in three dimensions: width (W), height (H), and depth (D).

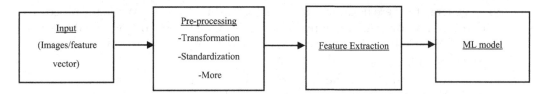

FIGURE 2.8
Image classification using ML.

The first layer in CNN starts with a convolutional layer that learns basic features (lines, edges, etc.); the next convolutional layer is responsible for learning complex features (circles, squares, and so on). Similarly, further stacked convolutional layers (if any) learn even more complex features (such as facial parts, complex contours, and so on) [6].

Figure 2.8 shows the steps of a classification model using machine learning techniques. The image features must be manually extracted to be fed into a machine learning system (e.g., SVM). The manual work of feature extraction and classification can be replaced by MLP or CNN; see Figure 2.9.

A basic CNN architecture with series of layers works in this manner:

INPUT => CONV => ACT => POOLING => CONV => ACT
=> POOL => FC => ACT(SOFTMAX MOSTLY)

The components of CNN are:

- Convolutional layer (CONV): This layer works similarly to the feature detector window by sliding over the image (pixel by pixel) with some fixed size and step; to mine some significant features for object identification in the respective image.

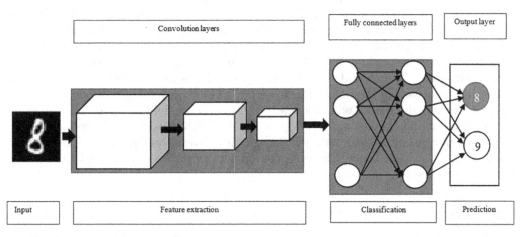

FIGURE 2.9
General architecture of CNN.

So, in general, they are used for feature extraction and learning. While the process is intuitive and powerful, repetitive stacking and the use of convolutional layers increase network dimensionality and space-time complexities. This is when pooling or subsampling comes to the rescue.

- Activation function (ACT): They convert the linear output into a nonlinear form. It is embedded after every perceptron, and it decides the activation of that neuron.

- Pooling layer (POOL): Pooling reduces the parameters given to the next layer, which results in a reduction in network size. The process of parameters reduction resizes its input using a summary statistics function like maximum or average.

- Fully connected layer (FC): FC layer is the normal dense layer that is a stack of neurons. It flattens the 2D grid of multiple features into a single 1D grid (a long tube) of values. These layers are responsible for learning and performing the classification task from the trained features.

- Batch normalization (BN): It is common practice to perform normalization before feeding the training data to the input layer; doing so benefits the model training and results. This can be done for each or a few selected layers of the neural network for better feature extraction and in turn increasing the training speed and network flexibility. The process is called batch normalization, where batch refers to the collection of parameters in a specific layer [6].

- Dropout layer (DO): This is an additional layer used to avoid the scenario of overfitting. Overfitting in learning from the training dataset occurs when the model fits the data but does not learn its features.

2.5 Image Classification Using CNN

Image classification is a technique used to classify the object(s) in an image into its respective class(es). There are mainly two types of image classification: multi-class and multi-label. A single class object is associated with an image in a multi-class classifier. For example, with a multi-class classifier, the classification of animals present in an image will result in whether it is a dog, a cow, or a cat. The subclass of the multi-class classifier is the binary classifier, where the CNN model differentiates only between two classes, such as cat or dog. The second type is the multi-label classifier, where the model has to label multiple objects in the image. For example, if there is an image containing several types of animals, the model will label each of them. In the field of image classification, a lot of research has been done on improving the CNN model and introducing new techniques like inception, residuals, etc. This section covers some popular CNN models. It also includes a walk-through of the development of CNNs from LeNet to AlexNet, VGGNet, and ResNet.

2.5.1 LeNet-5

LeNet is the first pioneering CNN proposed by Y. LeCun et al. [7] in 1998. This architecture was developed for textual data that is optical character recognition (OCR). The LeNet-5

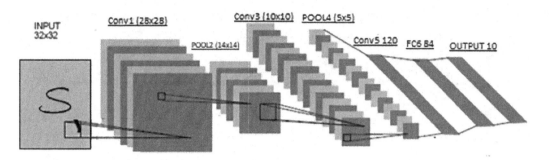

FIGURE 2.10
The LeNet-5 architecture.

architecture is straightforward with the essential CNN components: convolutional, sub-sampling or pooling and fully connected layers. Figure 2.10 depicts the model, consisting of five layers: three convolutional and two fully connected (hence the name "LeNet-5"). The model used the tanh activation function as it was considered that it would give better convergence than the sigmoid function. The LeNet-5 as a series of layers is as follows:

INPUT IMAGE => CONV 1 => TANH => POOL 2 => CONV 3 => TANH => POOL 4 => CONV 5 => TANH => FULLY CONNECTED 6 => SOFTMAX

2.5.2 AlexNet

LeNet performs well for the simple dataset like MNIST, where images are in grayscale and the number of classes is limited, ten in the case of the Modified National Institute of Standards and Technology (MNIST) dataset. To build deeper networks, the AlexNet model was proposed by A. Krzyzewski et al.[8], the winner of the ILSVRC Image Classification competition in 2012. The model was later published in 2017 with the title "Deep Convolutional Neural Networks with ImageNet Classification." 1.2 million images with high resolution from the ImageNet dataset were used to train the model, which was then divided into 1,000 categories.

This pioneering study on "deep" convolutional networks for computer vision sparked a storm of interest among researchers and practitioners alike. There are five convolution layers and three completely connected layers in the architecture, as depicted in Figure 2.11. This is how it looks:

- Five convolutional layers of kernel size 11 × 11 in Covn1, 5 × 5 in Conv2, and 3 × 3 in Conv3, Conv4 and Conv5.
- Max-pooling layer that performs maximum summary statistics function.
- Dropout layers (DO) to avoid overfitting.
- In the hidden layers, ReLU is utilized as an activation function, and in the output layer, softmax is used. The series of layers of AlexNet is as follows:

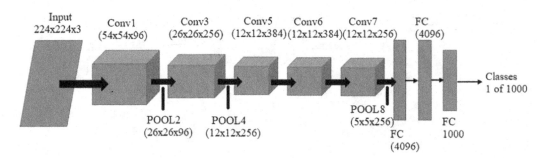

FIGURE 2.11
AlexNet architecture.

INPUT IMAGE => CONV 1 => POOL 2 => CONV 3 => POOL 4 => CONV 5 => CONV 6 => CONV 7 => POOL 8 => FULLY CONNECTED 9 => FULLY CONNECTED 10 => SOFTMAX

2.5.3 VGGNet

VGGNet was developed by the Visual Geometry Group at Oxford University in 2014, which is why it was named VGG [9]. It is a deeper convolutional neural network with more convolutional, pooling and dense layers. VGGNet is popular in two architectures: VGG16 and VGG19.

- VGG16 consists of sixteen weight layers: thirteen convolutional layers and three fully connected layers. The model is very simple and easy to understand. All convolutional layers are of size 3 × 3 and pooling layers are 2 × 2. The idea behind a small-sized kernel was to extract more fine features from the image, Figure 2.12.
- VGG19 has sixteen convolutional layers, five max-pooling layers, three fully connected layers, and a softmax layer.

2.5.4 Inception and GoogLeNet

The suggested deep CNN, called Inception, achieved state-of-the-art classification and detection efficiency in the ImageNet Large Scale Visual Recognition Challenge 2014

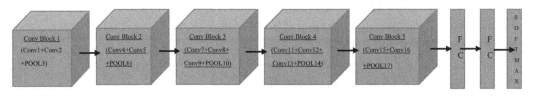

FIGURE 2.12
VGG16 architecture.

(ILSVRC14) [10]. By including the configuration into the model, the researchers increased the depth of the network while keeping the processing budget constant. GoogLeNet, a 22-layer deep network, was the model utilized in the ILSVRC14 proposal.

2.5.4.1 Inception Module

These are the little components that stack themselves on each other and form the Inception Network. A single Inception module is a combination of multiple convolutional layers aligned parallel to each other. See the Figure 2.13 for its complete architecture.

The input to these modules is the output from the previous modules. It is more computationally efficient to use the Inception module solely at the higher layers, leaving the lower layers alone, like in standard convolutional neural networks. The Inception modules use a 1 × 1 convolution to calculate the deduction before the expensive 3 × 3 and 5v5 convolutions. In addition to reducing feature dimensions and therefore being used for computation, 1 × 1 convolutions also use rectified linear activation and serve a dual purpose for the model.

Figure 2.14 shows that GoogLeNet contains nine inception modules in total, with a maximum pooling layer appended after each block to reduce dimensions. Let's divide GoogLeNet into three sub parts:

1. Similar to LeNet and AlexNet model which contains multiple convolutional layers and pooling layer connected in series.
2. Inception module: 9 inception modules (2 inception modules + 1 pooling layer + 5 inception modules + 1 pooling layer + 2 inception modules).
3. Classifier: fully connected output layer with softmax layer.

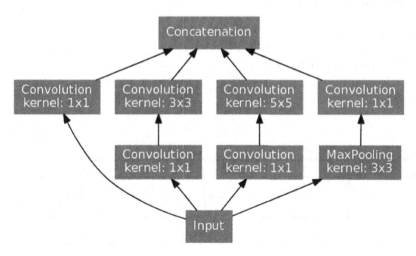

FIGURE 2.13
Inception module.

FIGURE 2.14
GoogLeNet model.

2.5.5 ResNet

ResNet, also known as Residual Network, was developed by one of the Microsoft Research Teams in 2015 [11]. Implementing deep NN with more layers like 50,100 may result in an overfitting problem. It may also cause a vanishing gradient problem. This problem can be solved using skip connections. Residuals are essentially skipped connections that allow activation from one layer of a NN to serve a deeper layer of the neural network. The idea behind ResNet is the implementation of residual blocks in a neural network to construct a deeper network. The main features of ResNet are:

1. It has residuals with skip connections
2. It has heavy batch normalization for hidden layers

2.5.5.1 Residual Block

A residual block consists of two paths; see Figures 2.13 and 2.15:

1. Shortcut path (x): It is a skip connection from one earlier layer to another later layer, as shown in Figure 2.13.
2. Main path f(x): It is the regular path consisting of convolutional and activation layers. In the case of the residual block, it has 3 Conv layers (1×1, 3×3, 1×1) and a batch normalization layer.

Components of ResNet50 are:

- One 7×7 convolutional layer
- Three Residual blocks, each having (1×1, 3×3, & 1×1) Conv layers = Total 9 Conv layers

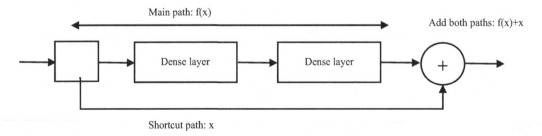

FIGURE 2.15
Residual block.

- Four Residual blocks, each having (1×1, 3×3, & 1×1) Conv layers = Total 12 Conv layers
- Six Residual blocks, each having (1×1, 3×3, & 1×1) Conv layers = Total 18 Conv layers
- Three Residual blocks, each having (1×1, 3×3, & 1×1) Conv layers = Total 9 Conv layers
- One fully connected softmax layer

2.6 Transfer Learning

Let us consider: a new model needs to be trained for the classification of different wild animals. Instead of training the neural network from scratch, the initial layers for feature extraction and learning might be replaced with a pre-trained model with similar applications. Many pre-trained models over various datasets are available open source over the web.

Building a large and deep neural network right from the start can take a lot of time, space and computation. This can be a tedious task and may take a few weeks or even months with high computational requirements. It is better to use the result of pre-trained models available from researchers over the years.

Transfer learning is the technique of transferring knowledge learned from a large dataset to a new similar application for some different dataset [12]. The performance of a network is related to the size and quality of the dataset in which it is trained. The lack of sufficient data may lead to under fit network. Using transfer learning, one can get better performance even for a smaller dataset.

2.6.1 Need for Transfer Learning

To attain great performance, deep neural networks require a significant amount of data. In general, only a few researchers train their neural networks from scratch because of the following problems: dataset problem and high computational requirements.

Dataset problem: Training a network from the start necessitates a massive amount of data in the dataset. Due to a lack of appropriate data for the deep network, the situation may not recover and will have a negative influence on performance. It is also not feasible in terms of cost and time to manually collect and process large amounts of data to create a dataset.

Computational requirements: Even if one can get a vast dataset or create it manually, the next problem of training a large network with lots of layers will require a highly configurable system with multiple GPUs. Training can be computationally expensive and requires weeks. The problem will not end here; training a deep NN necessitates many hyper parameter adjustments to improve its performance which will increase the time requirement.

2.6.2 Transfer Learning Approaches

Popular transfer learning approaches are: classifier, feature extractor, and fine tuning. All three can be effective and save computational cost and time for new computer vision applications. However, some experiments should be done to check their effectiveness for new deeper networks.

2.6.2.1 Pre-trained Network as a Classifier

In this approach, as shown in Figure 2.16, one can freeze the parameters of the pre-trained model and train only those parameters which are associated with the requirements of the new deep neural network. In Figure 2.16, the pre-trained classifier is used to classify the input image and the softmax layer is trained with the possible classes of the output according to the application of the model. This approach, where one can use someone else's pre-trained model to predict the different collections of images, will give good performance even with small datasets.

2.6.2.2 Pre-trained Network as a Feature Extractor

If a big dataset is available, transfer learning can also be utilized for feature knowledge sharing. The pre-trained model's initial layers can be frozen for feature extraction, and the remaining layers can be trained as classifiers. The classifier, as well as the softmax layer, needs to be trained for the new classification application; refer to Figure 2.17.

2.6.2.3 Fine Tuning

The above-mentioned methods of transfer learning can be used only if the target neural network has the same application as the pre-trained model, which is being used as a feature extractor or classifier. In the case of the target domain not belonging to an available pre-trained neural network, transfer learning can be used to extract the correct feature maps

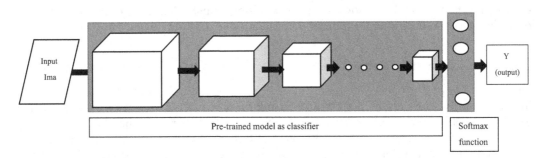

FIGURE 2.16
Classifier (pre-trained network).

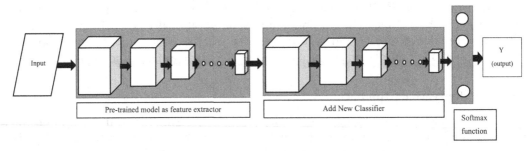

FIGURE 2.17
Feature extractor (pre-trained network).

from the already available pre-trained networks and fine-tune them as per the requirements of the target neural network.

2.7 Object Detection

Figure 2.18(a) shows a horse with a bounding box around it and the object is predicted as "horse." A model may be responsible for predicting that the image supplied is a "horse" in an object classification. The problem discussed in this section is the classification model with object localization. That is, the algorithm is responsible not only for predicting the object class but also for constructing a bounding box around the horse in the input image, or for drawing a red rectangle. The detection approach, on the other side, can handle a large number of objects. In fact, as illustrated in Figure 2.18(b), it may contain many objects from various categories within a single image. Object detection in computer vision is to find the coordinates of objects using bounding boxes as well as predict the respective class.

Object detection = object localization + object class prediction

2.7.1 Object Localization

Object localization is the process of drawing a bounding box (BB) nearby the located object. The object detector returns the coordinates for drawing bounding boxes around localized objects in an image.

2.7.1.1 Sliding Window Detection

The sliding window method is used to localize the object in the image. It is a rectangular box of a certain width (W) and height (H), also known as strides. It slides across the image. The image classifier checks whether each area within the window contains an object of interest. In this technique, a different-sized window moves over the image after each cycle to fit the object of interest inside the window. The computational cost of the sliding window detection approach is extremely high. Cropping out many square zones and running each one independently demands a lot of processing power. Moreover, if the window size

(a) (b)

FIGURE 2.18
(a) Classification and localization of horse in the image; (b) Object detection (cat, dog, and duck).

is large with a large step size to move the window, the number of windows necessary to check for categorization reduces, but performance may suffer. When the window size is small, however, a large number of regions are processed through the image classifier, resulting in a high computational cost. Fortunately, there is a reasonable answer to this dilemma. In particular, object detection with sliding windows can be done more effectively with convolutional neural networks.

2.7.1.2 Bounding Box Prediction

Although the convolutional sliding window approach is computationally efficient, it still has the issue of not producing the most precise bounding boxes. The bounding box produces a collection of tuples (x, y, w, h), where (x, y) are the coordinates of the center point, w is the box's width, and h is its height. Therefore, instead of making a square, it forms a rectangle with increased horizontal coverage.

2.7.2 Components of Object Detection Frameworks

Before moving on to the R-CNN family and advanced object detection models such as YOLO, here is a brief overview of the main components of the object detection framework:

- **Region proposal**
  Deep learning models scan each input and identify regions that may contain an object of interest. Further, separate bounding boxes create around and assign a prediction score to each of them. Based on the threshold value assigned prediction score helps to decide the focused and related objects in the image. The bounding boxes with higher scores are then processed by the next layer of the network using a selective search algorithm.

- **Feature extraction and network prediction**
  In this, visual features are extracted from the area below each bounding box in order to classify the object into its predicted class. It basically uses a classification or prediction model in which the network analyses all the regions within the bounding box that may contain the object with the higher score. The outputs of this component are bounding boxes with values (x, y, w, h) and class (probability of each class).

- **Non-maximum suppression (NMS)**
  This avoids having multiple bounding boxes on the same object. The NMS looks at all the bounding boxes nearby the identical object and checks for the box having maximum prediction probability, and it suppresses or removes the other boxes.

Following are the various cases for NMS:

- If the bounding box's prediction probability is less than the confidence threshold (default is 0.5), the NMS suppresses the box. From the remaining bounding boxes, it selects the bounding box with the maximum prediction probability.
- If two bounding boxes (BB) overlap each other and predict the same class, then the average of both is considered.
- If the IoU value between the bounding boxes is lesser than a set of confidence threshold (default is 0.5), then it will be suppressed.

2.8 Region-Based Convolutional Neural Networks (R-CNNs)

The region-based convolutional network (R-CNN) family is a recent dominant approach used for object detection. Fast-RCNN and Faster-RCNN are part of the R-CNN family. The evolution of R-CNNs, as well as their architectures, are covered in this section.

2.8.1 R-CNN

It is an object detection technique that uses a Region-based CNN approach to implement selective search with neural networks. It is a very early object detection algorithm that produces good results in a big dataset for object detection and localization. In the formative work of Girshick et al. [13], to obtain 2,000 region proposals per image, it was recommended to use an external object proposal, also known as a region proposal generator. From these region recommendations, CNN is used for features retrieval and SVM for identification purposes. Figure 2.19 depicts a model of R-CNN.

Components of the R-CNN model:

1. Input image: The model takes an image as an input.
2. Extract region proposals: The model then extracts region proposal using the selective search method. The selective search method iteratively group regions based on common features that might contain objects and then select around 2K region proposal for further processing.
3. Feature extraction from extracted region proposals: A pre-trained CNN model (e.g., AlexNet, ResNet) is then used to extract the features from these ~2K region proposals. In order to adjust each region proposal to fit as the input to the R-CNN, change the image dimensions in that region to fit with the CNN using dilation and wrapping.
4. Classification module: Based on the features retrieved in the preceding stage, a support vector machine (SVM) is used to classify the spotted objects. SVM classifier has a high recall rate but low precision. Therefore, it cannot be utilized as a stand-alone object detector, but it can be implemented as the first stage in a detection pipeline.

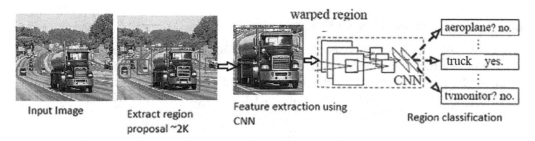

FIGURE 2.19
R-CNN model.

5. Localization module: Predict the bounding box coordinates that will be drawn over the localized object. This process is also known as regression problem to draw a bounding box. This step produces four parameters (x, y, w, h):

(x, y) is the origin coordinates, and (w, h) are the box's width and height.

The R-CNN approach has the advantage of being able to apply any CNN for feature extraction. The performance of CNNs can be improved by implementing more advanced convolutional architecture, which may result in increased mean average precision.

The basic R-CNN approach has its own set of issues. First, it is not very fast and does redundant computations. Finding regions using selective search can be very slow, and running each region of interest, up to 2,000 of them, via CNN can be slow and computationally expensive as well. Even for overlapped proposal regions, all features are computed independently. The second is that object proposals must be rescaled to a set resolution and aspect ratio. The third point is the hypothesis generation's reliance on an external algorithm. Finally, there is a highly difficult training process with a large file system capacity. The issue here is that the necessity to extract and save features generated by the CNN model for succeeding SVM training requires a large amount of memory.

2.8.2 Fast R-CNN

Ross Girshick, in 2015 [14], proposed an updated architecture known as Fast R-CNN to improve the base R-CNN model. Speed and memory concerns were somewhat resolved in Fast R-CNN by eliminating the costly selective search algorithm and some exciting architectural alterations. The basic R-CNN model has been rapidly improved. This section covers the transformations introduced in R-CNN to the Fast R-CNN model. Apart from the Fast CNN making use of CNN, the way the region proposal works is slightly different. Let's first examine the architecture changes in Fast R-CNN.

1. In Fast-RCNN, the CNN feature extractor is first applied to the entire input image, and then the regions are proposed. In this way, instead of 2,000 ConvNets in 2,000 overlapping regions, only one ConvNet operates over the entire image.

2. ConvNet is also responsible for the categorization portion of the process, in addition to feature extraction. In this, the traditional SVM machine learning algorithm was replaced with a softmax layer. By this, there is only one model to perform both tasks: feature extraction and object classification.

2.8.2.1 Components of Fast R-CNN

The Fast R-CNN architecture is depicted in Figure 2.20. A series of region proposals is sent to the network together with an input image. The red box on the input image is a visual example of a region proposal for this image. Instead of applying selective search for the generation of numerous region proposals for each image, Fast R-CNN accepts them as inputs and does not generate them.

- ConvNet: For feature extraction, to process the entire image, the network employs a combination of convolution (Conv) and max-pooling layers. Here, a feature map is a collection of extracted features.

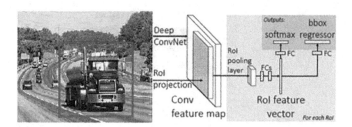

FIGURE 2.20
Fast R-CNN model.

- ROI pooling layer: This is a simplified illustration of a spatial pyramid pooling (SPP) layer with only one pyramidal layer. It retrieves feature vectors in a fixed-size window from the convolutional feature map and feeds them to fully connected (FC) layers. It uses a max-pooling layer to generate a smaller feature map with a fixed spatial extent HxW (*height × width*) for the features of any valid ROI.
- Output layers: It has two parallel output layers:
- The first output uses a fully connected layer to classify the image, followed by a softmax classifier.
- The second output utilizes multi-task training for simultaneous training of the classifier and irregular regression. The bounding box size and location coordinates for the classified object are the outputs.

The classic bag of Visual Words technique yielded the spatial pyramid pooling (SPP). A spatial pyramid is built on top of the ROI. The first level of the pyramid is an ROI itself. On the second level, the region is divided into four cells by two-by-two grids. The region is divided into sixteen cells on a four-by-four grid at the third level. Each cell is subjected to average pooling. As a result, if the final convolutional layer contains 256 maps, pooling in each cell produces a vector of length 256. All of the cells' feature vectors are merged and fed into the fully convolutional layer as input. Finally, for input images of various resolutions, we construct a fixed-length feature map.

2.8.3 Faster R-CNN

The model was proposed by Shaoqing Ren et al. [15] as an extension of Fast R-CNN and the third addition to the R-CNN family in 2016. A ConvNet produces a convolution feature map from the input image like Fast R-CNN. A region proposal network includes here to find ROI in the region proposal feature map.

Afterward, the respective image is classified inside the proposed region, and the bounding box is predicted.

The main components of Faster R-CNN are:

1. Region proposal network (RPN): When a fully convolutional network uses convolution, it is called an RPN. The RPN is built to be trained from start to finish producing high-quality field resolutions, which the R-CNN then utilizes for detection. The output of the RPN is the predicted object bound and the object score for each pixel.
2. Fast R-CNN detector: A pre-trained Fast R-CNN model used as an object detector.

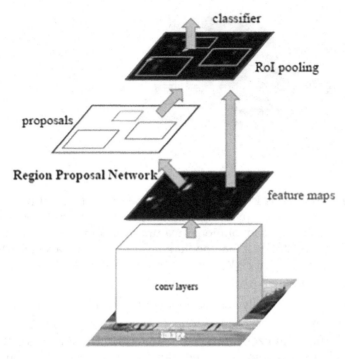

FIGURE 2.21
Faster R-CNN component architecture.

As illustrated in Figure 2.21, faster R-CNN passes the entire image through a ConvNet and generates feature maps using the pre-trained CNN model. The resulting feature maps are then sent in tandem to the RPN and ROI pooling layers. After that, the output is sent to the two FC layers. The object classifier categorizes the object according to its expected class, and the bounding box regression model predicts the coordinates for object localization.

2.8.4 YOLO Algorithm

YOLO stands for "you only look once." The YOLO family is also a collection of different objected detection frameworks that evolved and improved over the years. The members of the YOLO family are YOLOv1 (2016) [16], YOLOv2 or YOLO9000 (2016) [16], and YOLOv3 (2018) [17]. YOLOv1 is a single-detection network that integrates the two main components: detector and classifier. YOLO9000 is able to detect 9,000 objects. YOLOv3 is a deeper network than previously published versions and has better results. This section includes a description of how the YOLO model works.

Components of the YOLO family:
The YOLO model first divides the input image into SxS cells. Object detection in YOLO uses the ground truth (GT) box, sometimes known as the anchor box. If the center point of the GT box falls into a cell, then the object may be present in that particular cell. Every grid cell has the possibility of predicting B bounding boxes along with their probability prediction score. The steps are as follows:

  1. Predicting coordinates for B bounding boxes (bx, by, bw, bh).

2. Objectness score or probability prediction score (P_0) is the probability that a cell contains an object.

3. A classifier is used to predict the class to which the object belongs.

2.8.5 YOLOv1 Object Detection Model

The YOLOv1 architecture is a simple object detection architecture and easy to understand. It contains a single NN architecture that combines object detection and classification into a single end-to-end network. It is inspired by Inception and GoogLeNet for feature extraction. Here, YOLO uses a 1 × 1 reduction module instead of inception, followed by a 3 × 3 convolutional layer.

2.8.6 YOLO9000 Object Detection Model

It is capable of detecting 9,000 objects. It is a thirty-layer architecture (Darknet-19) with nineteen convolutional layers with additional eleven layers for object detection.

2.8.7 YOLOv3 Object Detection Model

YOLOv3 uses Darknet-53, shown in Figure 2.22. It is a 53-layer architecture trained over the ImageNet dataset. It is a very deep neural network. It has additional 53 layers for object detections (a total of 106 fully connected layers)

	Type	Filters	Size	Output
	Convolutional	32	3 × 3	256 × 256
	Convolutional	64	3 × 3 / 2	128 × 128
1×	Convolutional	32	1 × 1	
	Convolutional	64	3 × 3	
	Residual			128 × 128
	Convolutional	128	3 × 3 / 2	64 × 64
2×	Convolutional	64	1 × 1	
	Convolutional	128	3 × 3	
	Residual			64 × 64
	Convolutional	256	3 × 3 / 2	32 × 32
8×	Convolutional	128	1 × 1	
	Convolutional	256	3 × 3	
	Residual			32 × 32
	Convolutional	512	3 × 3 / 2	16 × 16
8×	Convolutional	256	1 × 1	
	Convolutional	512	3 × 3	
	Residual			16 × 16
	Convolutional	1024	3 × 3 / 2	8 × 8
4×	Convolutional	512	1 × 1	
	Convolutional	1024	3 × 3	
	Residual			8 × 8
	Avgpool		Global	
	Connected		1000	
	Softmax			

FIGURE 2.22
Darknet-53 [17].

2.9 Computer Vision Application Areas

The detection of objects using deep neural networks is an active and vast area of research in computer science. Research into object classification, localization, detection and recognition has been going on for the past few decades and researchers are still working on improving the performance of various deep neural network models. Some of the application areas are discussed in this section:

- **Autonomous vehicles:** One very popular application and expectation from artificial intelligence is full automation: no human intervention. When one talks about transportation, the first and foremost application that is most hyped among people is self-driving or autonomous vehicles. Currently, there are many possibilities to explore a serious amount of research and progress within the industry. Tesla cars and Google driverless cars are leading the industry with innovation and their application.

- **Autonomous drones:** Similar to the above scenario, automation tools are highly desirable and expected from AI. Automation of drone flight and functionality is also a potential application for so many tasks. Although the drone has some sort of AI In-built but fully automatic and collaboration with environmental cues is still an area of active research.

- **Crop insect detection:** Farming is an essential activity for our existence worldwide. It is very common for farmers and crops to become ruined as crop worms feed on them and destroy the entire crop. In many parts of Africa and other countries, AI software is designed to detect insects near crops that can potentially damage the crop. Detection of such an object (pest) alerts farmers and allows them to take necessary action before any significant damage occurs.

- **Video surveillance:** This application is becoming very common in all walks of life. This is on demand area of research for maintaining security in a particular area. Taking reference from a multi-national company, video surveillance is vital to the management and monitoring of employees and workforce. To manage power consumption by the office, many companies use visual wake-up words as triggers to turn the lights on and off. So that, when no person is visible in the picture, the light is turned off and electricity is saved.

- **Face detection and recognition:** This is just one of the many applications related to the face. Face detection specifically focuses on determining whether someone's face is present in the image. It does not see who is there, it sees whether a face is present and, if yes, where exactly in the image. Face recognition: On the other hand, face recognition is one step further from face detection. It focuses on whose face is present; recognition of that face is the end target for the model. It comprises two subtypes: face identification and verification. It also has some other applications like:
  - Sentiment analysis through facial expressions
  - Person identification and authentication
  - Age and gender prediction

- **Robot path planning:** For decades, robotics researchers have been working on object detection. It can aid in the detection of obstacles in the path of autonomous robots.

References

1. K. Gurney, "An introduction to neural networks," CRC Press, 2018, p. 234.
2. M. Mishra and M. Srivastava, "A view of Artificial Neural Network," *2014 Int. Conf. Adv. Eng. Technol. Res. ICAETR 2014*, 2014, DOI: 10.1109/ICAETR.2014.7012785.
3. C. Van Der Malsburg, "Frank Rosenblatt: Principles of Neurodynamics: Perceptrons and the Theory of Brain Mechanisms," *Brain Theory*, 245–248, 1986, DOI: 10.1007/978-3-642-70911-1_20.
4. J. Singh and R. Banerjee, "A study on single and multi-layer perceptron neural network," *Proc. 3rd Int. Conf. Comput. Methodol. Commun. ICCMC 2019*, pp. 35–40, Mar 2019, DOI: 10.1109/ICCMC.2019.8819775.
5. B. Ding, H. Qian, and J. Zhou, "Activation functions and their characteristics in deep neural networks," *Proc. 30th Chinese Control Decis. Conf. CCDC 2018*, pp. 1836–1841, Jul 2018, DOI: 10.1109/CCDC.2018.8407425.
6. M. Elgendy, *Deep Learning for Vision Systems*. Manning Publications, 2020.
7. Y. LeCun, L. Bottou, Y. Bengio, and P. Haffner, "Gradient-based learning applied to document recognition," *Proc. IEEE*, vol. 86, no. 11, pp. 2278–2323, 1998, DOI: 10.1109/5.726791.
8. A. Krizhevsky, I. Sutskever, and G. E. Hinton, "ImageNet Classification with Deep Convolutional Neural Networks," *Commun. ACM*, vol. 60, no. 6, 2017, DOI: 10.1145/3065386.
9. K. Simonyan and A. Zisserman, "Very Deep Convolutional Networks For Large-Scale Image Recognition," ICLR, 2015, DOI: 10.48550/arXiv.1409.1556.
10. C. Szegedy et al., "Going Deeper with Convolutions," *Proc. IEEE Comput. Soc. Conf. Comput. Vis. Pattern Recognit.*, vol. 07-12-June-2015, pp. 1–9, Sep 2014.
11. K. He, X. Zhang, S. Ren, and J. Sun, "Deep Residual Learning for Image Recognition," *Proc. IEEE Comput. Soc. Conf. Comput. Vis. Pattern Recognit.*, vol. 2016-December, pp. 770–778, Dec. 2015.
12. F. Zhuang et al., "A Comprehensive Survey on Transfer Learning," *Proc. IEEE*, vol. 109, no. 1, pp. 43–76, Jan. 2021, DOI: 10.1109/JPROC.2020.3004555.
13. R. Girshick, J. Donahue, T. Darrell, J. Malik, U. C. Berkeley, and J. Malik, "Rich feature hierarchies for accurate object detection and semantic segmentation," *Proc. IEEE Comput. Soc. Conf. Comput. Vis. Pattern Recognit.*, vol. 1, p. 5000, 2014, DOI: 10.1109/CVPR.2014.81.
14. R. Girshick, "Fast R-CNN," *Proc. IEEE Int. Conf. Comput. Vis.*, vol. 2015 Inter, pp. 1440–1448, 2015, DOI: 10.1109/ICCV.2015.169.
15. S. Ren, K. He, R. Girshick, and J. Sun, "Faster R-CNN: Towards Real-Time Object Detection with Region Proposal Networks," *IEEE Trans. Pattern Anal. Mach. Intell.*, vol. 39, no. 6, pp. 1137–1149, 2017, DOI: 10.1109/TPAMI.2016.2577031.
16. J. Redmon, S. Divvala, R. Girshick, and A. Farhadi, "You Only Look Once: Unified, Real-Time Object Detection," *Proc. IEEE Comput. Soc. Conf. Comput. Vis. Pattern Recognit.*, vol. 2016 - December, pp. 779–788, Jun 2015.
17. J. Redmon and A. Farhadi, "YOLOv3: An Incremental Improvement," *arXiv*, Apr. 2018.

3

Real-Time Tracing and Alerting System for Vehicles and Children to Ensure Safety and Security, Using LabVIEW

R. Deepalakshmi

Velammal College of Engineering and Technology, Tamil Nadu, India

R. Vijayalakshmi

Velammal College of Engineering and Technology, Tamil Nadu, India

CONTENTS

3.1 Introduction

Safety and security are the most important aspects for students and management to prevent students from being abducted, getting lost, etc. In the existing passive tracking system [1], a hardware device is installed in the vehicle that stores location, speed, etc.

When the vehicle returns to the specified location, the device is removed and data transferred to the computer. It also includes an auto-download option that transfers data via radio link but is not real-time [2]. The passive system did not prove to be very useful in tracking vehicles to prevent accidents. A real-time tracking system was required to transmit the collected information about the vehicle at regular intervals or at least transmit the information when required by the monitoring station. In this system, a device is fixed in the vehicle which will be interlinked with LabVIEW to track the vehicle in real-time and locate it on the Google map and the alerting system included alerts the user about the vehicle location only. This does not concentrate on the passengers in the vehicle [3]. This led to the development of active systems. Hence this automatic vehicle tracking system will help you to track the particular target vehicle as well as the students in real time and also alerts the parents and school management at regular intervals of time by using software called LabVIEW for vehicle tracking and RFID for student tracking. The active system is developed for the real-time tracking of a particular target vehicle at regular intervals of time [4]. By using LabVIEW, it captures the image of the vehicle and starts tracking it. The vehicle information is always stored in the database, such as time, speed, and location, connected with Google Maps. The students who are traveling in the vehicle are also tracked using RFID. Each student contains a tag that has a unique code. RFID system uses an electromagnetic field to transmit data from RFID to the tracker [5,6]. RFID indeed provides accurate and real-time tracking data for fixed and mobile assets. The vehicle and student information are stored in the database. The alarming system send the message to the parents and the management by getting information from the database linked whenever required. It is of great concern, especially in metropolitan cities where there is much more traffic than in others which makes it difficult to expect our children to be at a specified location on time [7]. The main aim of this work is that the whole alarming unit will provide you with a way to ensure the safety of children by tracing locations and alerting on misfortunes and acknowledging successful departure at the specified location [8,9]. The assurance of safety is usually the most unknowing but still the essential one. So, this system enables you to keep this asset at your fingertips using only your basic essentials.

3.2 Scope of the Chapter

- To track the particular target vehicle using LabVIEW
- To guarantee the safety and security of the students using RFID
- To alert parents as well as management
    i. vehicle location information
    ii. diversions from regular routes
    iii. absence of the student at their stop
    iv. acknowledgment of student's arrival

3.3 System Requirements

3.3.1 Hardware Requirements

- Processor – Intel Dual core 2.0
- Motherboard – Intel DG41
- RAM – 2 GB
- HDD – 250 GB SATA
- Keyboard – MM Keyboard (USB)
- Mouse – Optical Mouse (USB)

3.3.2 Software Requirements

- Browser: Firefox, Edge, Chrome, etc.
- Operating System: Windows 10 with core i5 processor
- IDLE: LabVIEW version 18.0 (32-bit)

3.4 Real-Time Tracing and Alerting System Environment

This chapter proposes a system whose purpose is to work in a formless environment. In the formless environment, there is no false blue/green screen which gives improved system flexibility and movability but can make reliable division more difficult [10]. Limiting the target objects to saturated and characteristic colors enables them to be illustrated from the formless background, overcoming the limitation. Expanding the formless atmosphere with structured color is a negotiation that enables a segmentation algorithm to be used.

3.4.1 Image Acquisitions Module

The image capture is performed using a color video camera which produces a stream of RGB pixels. The area of search on the image is limited by using the general knowledge of a street perspective. The image is analyzed through a moving window and is measured by a detector of a different resolution multiple of 32×32. The camera monitors the vehicle to be tracked. A brief 5–10-second video is recorded in film format. To determine the exact location of the region of interest, run the dummy vehicle detection software on the system while on the road. This detects vehicles, and we notice that all the vehicles are located in a particular region. The video frame size is set to 620×340 camera and the angle of pixels. The tracked continuous video is given as an input to the software using LabVIEW. The software processes the entire video and converts it into image frames. This depends on the placement of the camera and the angle of view, which cannot be changed [11].

3.4.2 Object Identification Module

In this module, the tracking vehicle should be differentiated from the circumstantial environment. The elimination takes place so that the tracing vehicle can be easily identified by converting it into grayscale. The first frame is set as the background frame is selected [12]. In view of the recorded video, any frame is randomly selected as the object frame for object recognition which gives information about the shape and size of the object (Figures 3.1 and 3.2).

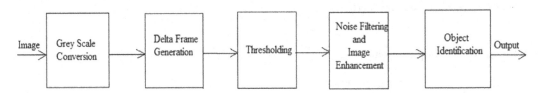

FIGURE 3.1
Flow of object recognition.

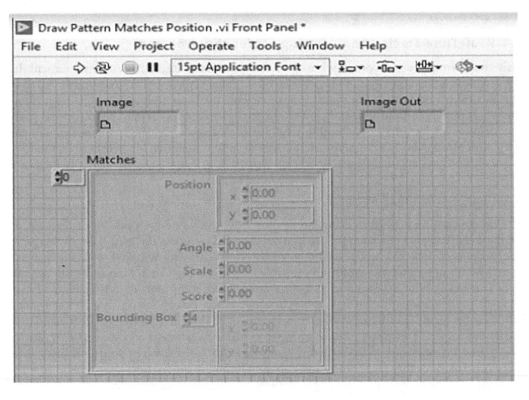

FIGURE 3.2
Front panel.

3.5 System Architecture

The domain of the project consists of two modules: The first module is vehicle tracking; this module tracks the target vehicle in real time by using the software LabVIEW and stores the location information and vehicle video and stores the collected information in a database which is interlinked with the software [13]. Locating the moving object is the process of vehicle tracking in time using a video camera. The system analyses the video frames and outputs the location of moving targets within the video frame. The school vehicle tracking system uses LabVIEW software to track the vehicle.

This is the vehicle-tracking algorithm, using the LabVIEW simulation tool:

Step 1: Capture and save the image of the vehicle using the LabVIEW simulation tool.
Step 2: Check if the captured vehicle is the same as the target vehicle.
   a. If YES, track the vehicle until the destination is reached.
   b. If NO, capture and check for the target vehicle until found.
Step 3: Once the target vehicle is found and tracked, check if the location information is available.
   a. If YES, store it in the database server.
   b. If NO, again continue with step 2b.
Step 4: Make sure the location information stored serves the needful and use them to track the vehicle in Google Maps.

The general flow of an object tracking algorithm is described as follows: (1) An arrival model is recognized based on the first information. (2) The arrival model is used to determine the object's location in the current frame. (3) Based on the tracking results for the current frame, an strategy is used to update the arrival process to allow it to adapt to changes in the object and the environment. The arrival process is to analyze the frames in order to estimate the signal limitations. These limitations describe the target location, which in turn helps to identify features like average speed, direction changes, total time in motion and form and scope of the target (Figures 3.3 and 3.4).

The second module is the student-tracking system. It recognizes the student who enters the vehicle at the source location by reading the tag held by the student using RFID (radio frequency identification), monitors the presence and absence of students and stores the information in the database [14]. By integrating this system with DBMS (database management system), a database can be maintained to monitor the speed and the location at which the nineteen vehicles had traveled. The alerting system can also be provided in case of any speed violation by the traveler. The speed limit can be set by the owner of the vehicle, and they can receive notifications through the mail in case the vehicle travels above the speed limit specified by the user (Figure 3.5).

The following is the algorithm for student tracking, using an RFID tag:

Step 1: When the student enters the bus, the sensor recognizes the RFID tag details, and the data is sent to the RFID reader.
Step 2: This reader decrypts the message and sends it to the attendance application.
Step 3: The attendance application compares the information with the database.

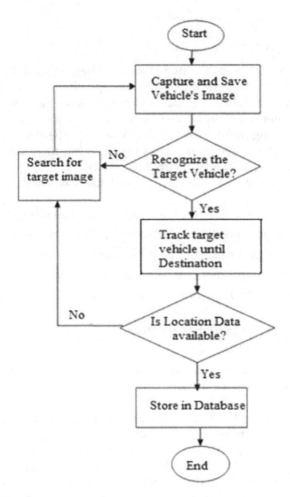

FIGURE 3.3
Vehicle-tracking system flow diagram.

Step 4: Checks if the data matches the database.

    a. If YES, attendance is marked, saved and acknowledged.

    b. If NO, absence is marked, and a notification is sent to the parents and alerts them.

Step 5: Parents are acknowledged on students reaching their destination.

3.6 Outline of Vehicle-tracking Processing

The basic working principle of a vehicle-tracking system is to track a specific target vehicle or other objects. The tracking device is able to relay information concerning the current location of the vehicle and its speed. Tracking systems consist of an electronic device usually installed in the vehicle.

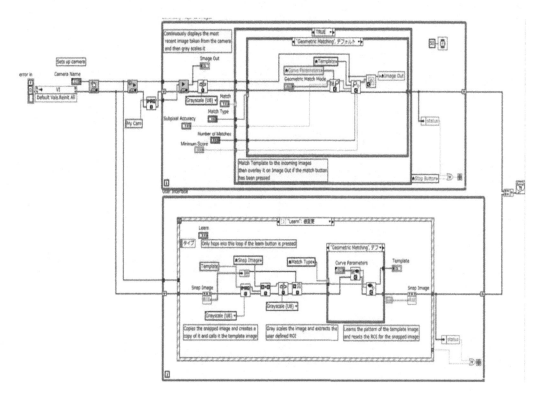

FIGURE 3.4
Block diagram of vehicle tracking.

The proposed method analyses the video frames one by one using LabVIEW and outputs the direction and location of the moving vehicle in the video frame. The algorithm takes one frame from the image and tracks the vehicle using pattern matching which analyzes the video frames in order to approximate the motion parameters (Figure 3.6).

3.7 Overview of RFID

In this system, a PIC16F877A microcontroller has been used. The system consists of three units, bus unit, school unit and parent unit. The bus unit consists of an RFID reader, fire sensors and SMS to alert messages to parents when their children board or leave the bus. A fire sensor will be placed within the bus unit to detect fire and issue alert messages by giving the location of the bus using IoT. The school unit consists of an RFID reader and GSM Module. The entire data in two units will be processed by using a PIC16F877A microcontroller. This processor has advantages like the total number of pins is 40, and there are 30 pins for input and outputs, 368 RAM bytes, 5MIPS CPU speed, and 8 channels of 10-bit ADC converter. In this system fire sensor is used to detect fire accidents. If any fire accident occurs, the alert message will be sent to the school unit and parents with the help of IoT and SMS. Each student unit consists of an individual RFID tag; with the help of the RFID tag, the parents and school units can receive an alert message.

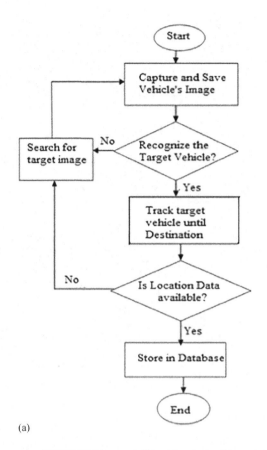

(a)

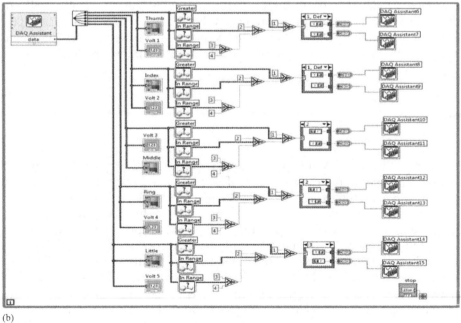

(b)

FIGURE 3.5

(a) Flowchart for student-tracking system; (b) Block diagram of student tracking.

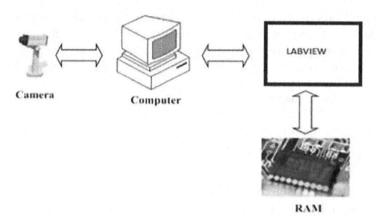

FIGURE 3.6
Layout of image-processing system.

The information of the RFID tag is read by the RFID reader. The reader transmits the corresponding information. The RFID tag is used to send an alert message like a person's location and speed of the bus to their respective parents. In this system, IoT is used to send the alert message to the parents if their respective child is getting on or off the bus with the help of RFID tag and reader. LCD stands for liquid crystal display, a flat panel display technology commonly used in TVs and computer monitors. It is also used in screens for mobile devices, such as laptops, tablets, and smartphones. The backlight in liquid crystal display provides an even light source behind the screen. This light is polarized, meaning only half of the light shines through to the liquid crystal layer. The liquid crystals are made up of a part solid, part liquid substance that can be "twisted" by applying an electrical voltage to them. They block the polarized light when they are off, but reflect red, green, or blue light when activated. ADC Power Supply Unit (commonly called a PSU) deriving power from the AC mains (line) supply performs a number of tasks: It changes (in most cases reduces) the level of supply to a value suitable for driving the load circuit. It produces a DC supply from the mains (or line) supply AC sine wave. It prevents any AC from appearing at the supply output. Power supplies in recent times have greatly improved in reliability but, because they have to handle considerably higher voltage currents than any or most of the circuitry they supply, they are often the most susceptible to failure of any part of an electronic system. GPS is a satellite navigation system used to determine the ground position of an object. Each GPS satellite broadcasts a message that includes the satellite's current position, orbit, and exact time. A GPS receiver combines the broadcasts from multiple satellites to calculate its exact position using a process called triangulation.

3.7.1 Implementation of RFID

There are various options available to ease the process of monitoring and tracking students. One tracking technology offers complete efficiency in the most cost-effective manner: RFID. In its simplest form, RFID student tracking is a way of mechanizing the organization and locating process of students. It works by loading an RFID tag with data and attaching it to a relevant student, and the information contains the data like student name, class, boarding point, and route number.

An RFID tag's continually lively radio waves read to capture the stored data like student name, class, boarding point, and route number. In the end, collecting it in a refined student tracking system where the data can be monitored and actioned. Automating student tracking and monitoring processes aims to end the highly error-prone methods of pen-and-paper and Excel spreadsheets.

Among other benefits such as:

- Tracking vehicle at any one time
- Removing human intervention
- Gathering data in real-time
- Improving student distinguishability
- Locating lost or missed students
- Exploiting accuracy of tracking

The student RFID tracing system requires the following tools:

- RFID tags
- An antenna
- An RFID reader
- A computer equipped with student-tracking software

The student tracking system consists of hardware modules, android modules, and web-based applications. The system is divided into three central units:

A. Vehicle-tracking module
B. Parent security and alert module
C. School module

A. Vehicle-tracking module
This module is designed for sensing the student when the student enters and exits the school vehicle and sends the data to the school module and to the parent security alert module.
This module consists of

a. radio frequency identification detection reader and tag
b. Global Positioning System module
c. switch
d. PIC16F877A microcontroller which is integrated with RFID reader tags

The reader is placed at the front door of the school vehicle, and the module works as both transmitter and receiver. The RFID tag is embedded in a student's ID card. And using the modem connected with the server is used for communication between the rest of the units. It is used for school vehicle tracking and alerting if the vehicle crosses the speed limit assigned. The server board consists of a button that is responsible for tracking the vehicle and finding whether it is met with any major trouble then it can use the button in the microcontroller which in turn sends the message to the server [15–18].

B. Parent security and alert module

The android application connected with the parent security and alert module has the parent sign-up option to log in with the mobile number registered in the student database maintained by the school to get the student's information and notifications automatically and able to track the school vehicle in which the student is assigned. If there is a chance that the parent does not have an android phone, then the information will be sent as a standard message to the parent's phone. In this, the drawback is that the parents will not be able to view the live tracking of the student's vehicle location.

C. School module

This module has a web-based application developed for a school database with an administrator who is responsible for all student details entries and is permitted to do all database activities like add, delete, update, and modify the details of the school vehicle, student details, parent details, vehicle routes, string and end information as required it has been saved on the server. And based on the GPS Module attached to each school vehicle, the school module can retrieve the current locations of the vehicle with the predefined route and schedule assigned to each stop from source to destination. The alert message is instantly generated and will be received by the school module whenever the vehicle crosses the assigned speed limit.

The speed limit can be set by the owner of the vehicle, and they can receive notifications through the mail in case the vehicle travels above the speed limit specified by the user (Figure 3.7).

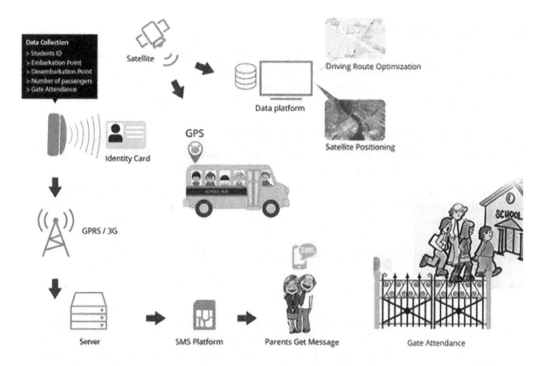

FIGURE 3.7
School unit.

3.8 Module

1. Vehicle tracking using LabVIEW
2. Student tracking using RFID
3. Alerting system

3.8.1 Vehicle Tracking Using LabVIEW

LabVIEW is used for automated manufacturing tests, validation, control and monitoring of a system. For real-time tracking of vehicles, LabVIEW captures the image of the target vehicle by using the in-built camera and then highlights the captured image using a rectangular red box. Once the LabVIEW diagram and the vehicle image get simulated, it starts tracking the vehicle until it reaches the destination. While tracking the vehicle, information like the vehicle's location and time information gets stored in the database.

It recognizes the student who enters the vehicle at the source location by reading the tag held by the student using RFID (radio frequency identification), monitors the presence and absence of students and stores the information in the database. By integrating this system with DBMS (database management system), a database can be maintained to monitor the speed and the location at which the nineteen vehicles had traveled. The alerting system can also be provided in case of any speed violation by the traveler. The speed limit can be set by the owner of the vehicle, and they can receive notifications through the mail in case the vehicle travels above the speed limit specified by the user (see Figure 3.8).

3.8.2 Student Tracking Using RFID

In this module, students get tracked by using RFID technology. Parents need a method or technology to monitor and track their children's safety during their travel to and from school. RFID is a data collection method that involves the automatic identification of objects using RFID tags, an antenna, an RFID reader, and a transceiver.

Step 1:   Wearable RFID tags are given to all students.

Step 2:   An RFID reader is installed in the vehicle. When the student enters the bus, it automatically scans the reader.

Step 3:   RFID identifies the tag number and stores the student information such as the student's source location, absence of students and the entering time the database.

The following steps describe RFID features and event handling:

- RFID generates a reader signal to imitate the instance of the tracking of a school vehicle.
- The communicated vehicle port number and location identification can be identified by its current IP address by the in-built methods.
- Once the students board the vehicle, their attendance will be noted by the RFID tag.
- The time between boarding and getting off will be calculated using the time gap.

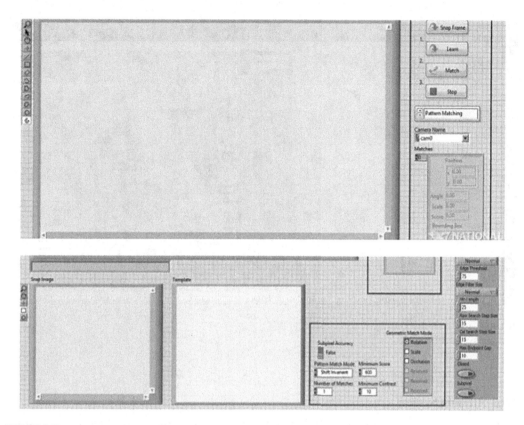

FIGURE 3.8
LabVIEW implementation.

The database defines and establishes the data of students, staff, and administrators (Figures 3.9 and 3.10).

Therefore, the system has the following functions:

i. New addition: The administrator uses a method called addition used to add an RFID reader for each classroom

ii. Display or Show: this shows the student's availability

iii. Update: stores the student's login details

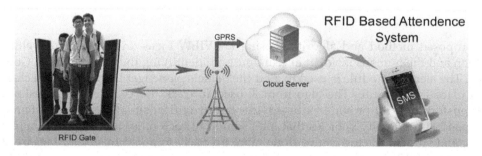

FIGURE 3.9
RFID-based student tracking.

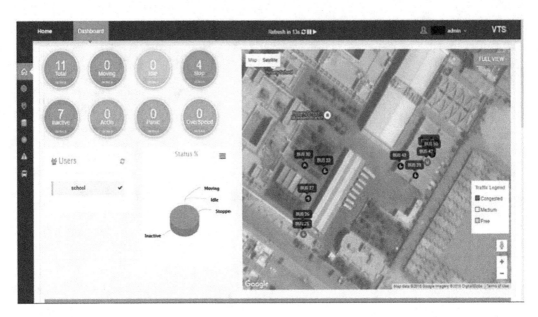

FIGURE 3.10
Tracking system of bus.

3.8.3 Alerting System

Collected information from the first module is stored in the vehicle database, and the information about the students is stored in the relevant ID. In the module, a solution is provided using RFID technology to track students journey to/from school, thus notifying the parents about the status of their children. This reads the children's RFID tag and automatically sends different generated messages to their parents [19]. The proposed solution has a set of capabilities, such as notifying the parents that the children have arrived at the school safely, and if any children are absent without authorized permission, then an immediate message is sent to their parents (Figure 3.11).

3.9 Conclusion

The proposed method and algorithm using LabVIEW focus on tracking the vehicle's position and the humans inside it in real time and identifying the location using Google Maps. The alerting module is used to alert the vehicle passenger if any direction change takes place in the route. The algorithm is used to track the vehicles, stop customers from misfortune, and provide extra services in any critical situations. This system can be expanded to identify vehicles that look similar to each other by using their number plates, and student tracing can be done more precisely to avoid misuse of tags by students (Figure 3.12).

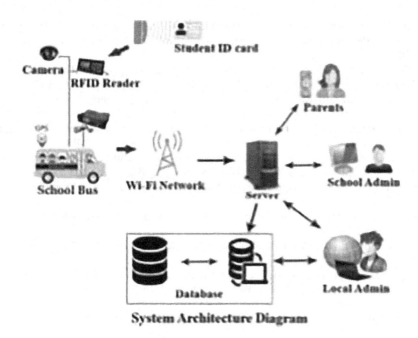

FIGURE 3.11
Alerting system.

FIGURE 3.12
Tracking of vehicle using LabView.

References

1. Akila Wajirakumara, Vehicle Tracking System Using GPS and SMS.
2. David Moore, A real-world system for human motion detection and tracking California Institute of Technology, dcm@acm.org, June 5, 2003.
3. GPSImages. http://www.gpsvehiclenavigation.com/GPS/images.php

4. S. Lee, G. Tewolde, and J. Kwon, "Design and implementation of vehicle tracking system using-GPS/GSM/GPRS technology and smart phone application." Internet of Things (WF-IoT), IEEE, 2014.
5. A.S. Dinkar and S.A. Shaikh, "Design and Implementation of Vehicle Tracking System Using GPS," *Journal of Information Engineering and Applications*, Vols. 1, 3, pp. 1–6, 2011.
6. S. Murugananham and P.R. Mukesh. "Real Time Web Based Vehicle Tracking Using GPS." *Journal of World Academy of Science, Engineering and Technology*, Vol. 61, pp. 91–99, 2010.
7. P.K. Harshadbhai, "Design of GPS and GSM Based Vehicle Location and Tracking System," *International Journal of Science and Research*, Vol. 2, no. 2, pp. 165–168, 2013.
8. D. Dao, C. Rizos, and J. Wang, "Location-based services: technical and business issues", *GPS Solutions* Vol. 6, pp. 169–178, 2002, doi: 10.18535/ijecs/v4i8.05.
9. S.S. Reddy, "Trip Tracker Application on Android," MSc Thesis, San Diego State University, USA, 2011.
10. T. Le-Tien, and V. Phung, "Routing and Tracking System for Mobile Vehicles in Large Area," IEEE, Vietnam, 2010.
11. T. Le-Tien, and V. Phung, Vietnam "Routing and Tracking System for Mobile Vehicles in Large Area," IEEE July 2009.
12. P. Chaiprapa, S. Kiattisin, and A. Leelasantitham, "A Real-Time GPS Vehicle Tracking System Displayed on a Google-Map-Based," Website. http://department.utcc.ac.th/research/images/stories/5207003.pdf
13. F. M. Franczyk, and J. D. Vanstone, "Vehicle warning system," Patent number: 7362239, 22 Apr 2008.
14. G. Xu, "GPS Theory, Algorithms and Applications" Springer, 2007.
15. S. Piot, Security over GPRS, Master of Science Mohd Roshmanizam Bin Hamad Rodzi, "An Enhancement of Vehicle Security Alarm System via SMS," Bachelor of Science in Data Communication and Networking, MARA University of Technology, 2006.
16. Car Security and Tracking System with Position, Route, and Speed Calculation, EDP Topic VG06, 2011.
17. Will Jenkins, Real-Time Vehicle Performance Monitoring with Data Integrity, Mississippi State University.
18. J.C. Coffey, J. Palmer, "Implementing Student Tracking Systems at Community Colleges," Washington, DC: American Association of Community and Junior Colleges, 1990.
19. S. Mayer, Impact of GPRS on the Signaling of a GSM-based Network, Institute of Communication Networks and Computer Engineering, University of Stuttgart.

4

Mobile Application-based Assistive System for Visually Impaired People: A Hassle-Free Shopping Support System

E. Ramanujam

National Institute of Technology Silchar, Assam, India

M. Manikandakumar

Thiagarajar College of Engineering, Tamil Nadu, India

CONTENTS

4.1 Introduction

The number of visually impaired (VI) people worldwide is increasing considerably, and may continue to increase [1]. In their report, the World Health Organization (WHO) stated that over 230 million people are visually impaired, 37 million of whom are blind [2].

DOI: 10.1201/9781003206736-4

Several factors can cause vision loss. The causes of impairment include eye damage, the brain's inability to accept and comprehend visual cues from the eyes, and so on. In most cases, the visual impairment can be caused by the common underlying disorders such as cataract formation, diabetic retinopathy, age-related macular degeneration, and increased intraocular pressure in the eyes, which leads to glaucoma [3].

Vision loss can strike at any time in one's life. In rare situations, visual impairment can be passed down from parent to child. In such circumstances, it can even show up during pregnancy or early childhood. Recently, genetic defects, developmental defects, retinitis pigmentosa, and other disorders are the common conditions for visual impairment. Children may have a partial or total disability or delayed impairment during their development, especially in gross and fine motor skills, reported in [4]. Adults who are visually challenged have difficulty finding work and need assistance to carry out their daily lives.

The most prevalent eye illnesses that cause blindness are:

- cataract,
- age-related macular degeneration (AMD),
- diabetic retinopathy, and
- glaucoma.

4.1.1 Cataracts

A cataract is defined as a cloudy environment surrounding the lens of the eye [5]. As a result of the cataract, the light exchange pathway necessary for vision is blocked. The lens of the eye is located behind the iris and pupil and appears to be made of transparent material. The role of this lens is to help focus the images into the retina, which is generally behind the back layer of the eye. This will transmit all the observed images to the brain. Cataracts will cause vision to be blurred and dimmed because the natural light cannot be transmitted correctly through the lens toward the retina. Due to this, the eye protein may stagnate near the lens, which gradually leads to the formation of cloudy environments termed cataracts of the eye. Vision that becomes clouded, blurry, foggy, or filmy is a typical problem [6]. Nearsightedness in the elderly is the other form of cataract. It changes the color perception, concerns such as glare from oncoming headlights while driving at night, and glare issues throughout the day. Cataracts come in various forms such as nuclear sclerotic, cortical, posterior subcapsular, anterior subcapsular, congenital, traumatic, radiation, lamellar or zonular, posterior and anterior polar, and Christmas tree cataracts.

4.1.2 Age-Related Macular Degeneration (AMD)

Age-related macular degeneration (AMD) is another primary reason (eye illness) for severe and permanent sight loss in adults [7]. It occurs when the macula, the little core section of the retina, wears down. The retina is the rear of the eye's light-sensing nerve tissue. There are two forms of age-related macular degeneration [8]. The first type is the dry form, which causes yellow deposits in the macula known as drusen. This may cause the eyesight to dull or distort, especially when reading. The light-sensitive cells in the macula become thinner and finally die as the illness worsens. The second variety is called wet form, and it occurs when blood vessels in the retina leak blood and fluid. The bleeding from these blood vessels usually forms a scar, resulting in irreversible loss of central vision. Smoking, high blood pressure or cholesterol, obesity, consuming a lot of saturated fat and females having light-colored eyes may be risk factors for AMD.

4.1.3 Diabetic Retinopathy

The eye illness or the eye disease diabetic retinopathy will mainly affect the very small blood vessels in the retina of persons who have diabetes [9]. Diabetes will cause the small blood veins in the retina to become weak; either it will break down, or it can become blocked. This is the most widespread and leading source of visual impairment and blindness.

4.1.4 Glaucoma

This is a category of disorders that are frequently linked to high pressure which is created inside the eye [10]. This disorder may produce high pressure to the eye and damage the optic nerves that convey information from the eye to the brain. The damage will be progressive, beginning with peripheral vision loss and progressing to central vision losses, which could lead to blindness.

The issues discussed above generate a great demand for assistive devices/services to support blind and VI individuals. Since a VI person's ability to focus cannot be corrected to a normal visual level, blind people face a greater number of challenges, such as navigating and roaming around places, searching for reading materials, arranging clothes, getting things independently, identifying routes, avoiding potential accidents, and so on.

Many popular physical tools or assistive techniques are available, like white canes and guide dogs [11–14]. These physical tools do not perceive and provide information like a person with normal vision. To help visually impaired people, researchers have created an assortment of portable and wearable assistive gadgets for carrying out their daily activities in indoor and outdoor environments. In addition, [14–17] have invented several technologies to address the lack of vision issue, which are grouped and displayed in Figure 4.1.

Among the devices listed, wearable devices for the blind have more significance in assisting the VI people in various scenarios. However, they are the much-underused form in the interaction by the VI people in real-time like other devices such as smartphones, bracelets, smartwatches and glasses [18]. Though various electronic devices are available, there is no special device to facilitate the VI in hassle-free shopping. The VI people suffer a lot from observing and carrying out routine tasks such as navigation at home, city and shopping in malls. For example, the visually impaired have particular difficulties purchasing products, reading the labels from the product, information about the product like expiry date, manufacturing date, maximum retail price (MRP), and vendor name. This may sometimes lead to bad decisions and consume more time on purchasing products.

There is plenty of analysis and work being done to look out for approaches for the up life for VI and partially sighted individuals in hassle-free shopping. Because of developments in recognition and reading technologies, smartphones, tablets, and smart glasses may become essential tools for VI people, as reported in [15,17].

This chapter proposes a mobile-based assistive product to identify the information about the product by scanning the barcode present in the product for hassle-free purchase. The implementation has been carried out in the controlled laboratory environment with a minimum of 2,113 products related to groceries, fashion items, bathing soaps, etc. This proposed assistive technology with state-of-the-art services and applications will make their shopping activities easier for the VI people.

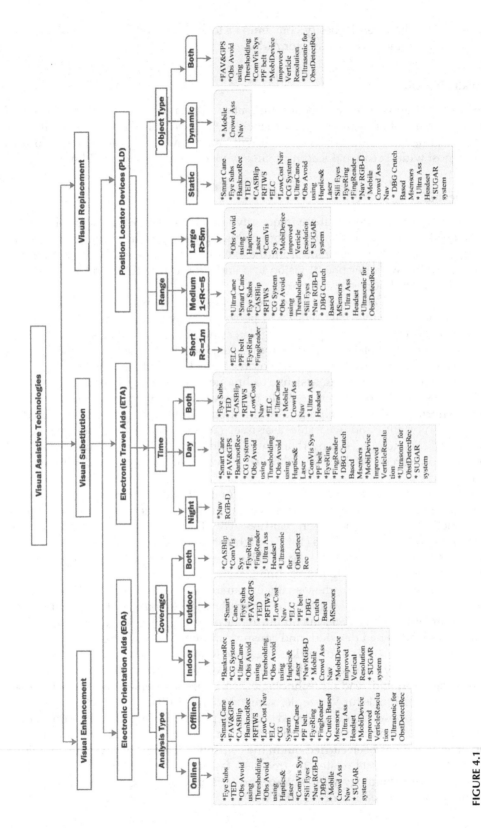

FIGURE 4.1
Categorization of devices developed for the welfare of visually impaired people [11].

The remaining chapter is sectioned as follows. The complete literature on assistive devices for the hassle-free purchase of products is discussed in section 2. Section 3 deals with the proposed methodology that includes pre-processing, barcode detection, barcode decoding, product specification database and voice-to-text conversion. Section 4 discusses the performance analysis of the proposed methodology with discussion and, finally, section 5 concludes the chapter.

4.2 Related Works

The section reviews the existing assistive product to help blind people purchase products in supermarkets or grocery stores. People suffered from various issues in the outside world, such as the purchase of products, roaming to or within the store, picking up the products from the racks/shelves, details about the product etc. Each visually impaired person may have a different sense of impairments, and their challenges will be according to the level. To resolve these issues, various assistive devices/techniques have been designed and developed by the industries and researchers and commercial products available in the market [19]. Assistive devices include low-level to high-level sensing devices attached to the trolley basket, wearable devices, and smartphone-based detection techniques. The major challenge in purchasing the products is identifying/finding the items in the store, price, and details about the products [20].

4.2.1 Item Identification

Persons with total loss of vision may identify any products through Braille to read an identification using the haptic sense [21]. In India, a smaller number of products use Braille and, unfortunately, many products do not have Braille labels; this makes them critical in the purchase of products. In some way, the VI/PVI person identifies the product by detecting the shape. However, many products have the same shape. For instance, cylindrical cases are common for most energy drinks. The cross-product design similarity clearly increases the challenges in purchasing and distinguishing the products among their types and brands.

Multiple techniques have been developed for identifying the products without Braille labels. Optical character recognition (OCR) is a technology that translates handwritten text (scanned form) into machine-readable text. Radio frequency identification (RFID) tags are also used to track RFID-tagged e-information decoded in the objects [22]. The extracted text can be provided to PVI through smartphones for easy access to product information. The research papers [23–25] used the concept of barcode scanning for the identification of products and to retrieve relevant information.

Despite the provision, the OCR and RFID have certain limitations. OCR supports only the language written in English and that too not the cursive written English language. RFID needs a well-known and created database and deployment of sensors to categorize the products, which is more costly [26]. This makes it difficult to purchase serendipitous shopping products by the PVI. To overcome these issues, this chapter investigates the critical understanding of barcode detection for the hassle-free purchase of the products by the PVI. Before reviewing various research works categorized for hassle-free shopping, the barcode and its characteristics are briefed.

4.2.2 Barcodes

Barcodes are classified into 1D and 2D barcodes based on the character set and are generally of four types: numeric, alphanumeric, GS1 AI encodable character set 82, and full ASCII [27]. A numeric character set includes numbers from 0 to 9, the alphanumeric sets consist of alphabets (A-Z) and numbers (0-9), whereas the G1 AI encodable set has some special characters in addition to alphanumeric sets. Finally, the full ASCII set holds any ASCII character from 0 to 127 values. The one-dimensional (1D) barcode has vertical black lines with white background of varying widths and gaps, resulting in a specific pattern as shown in Figure 4.2. The two-dimensional (2D) barcodes have two-dimensional square/rectangular patterns that are primarily black against a white background, as shown in Figure 4.2. For example, European Article Number (EAN) and Quick Response (QR) codes are 1-D and 2-D barcodes, respectively. The two-dimensional barcodes hold more data than 1D and thus have a larger character set.

During the last two decades, various research works have been proposed for the welfare of PVI to address their issues in hassle-free shopping. Almuzaini et al. (2019) have proposed a smartphone-based medication identification system for PVI people using near field communications (NFC) and barcode detection [26]. Using this system, PVI people can easily identify the medication without depending on any neighbors. Jethjarurach and Limpiyakom (2015) proposed a mobile-based product identification for blind people in the Thailand region to enhance the quality of their lifestyle and their ability to live independently [24]. This research work scans the barcodes using a mobile phone and preprocesses the images and identifies the products from the database. The information has been conveyed as a voice message through a text-to-speech synthesis tool integrated in the mobile phone. [23] has introduced a navigation system using a hybrid tag-based camera that uses a portable camera, computing device and barely hardcore content identification from barcodes for the welfare of PVI people.

[28] has utilized a smartphone to recognize QR codes for object identification in the environments. The system uses a mobile camera that scans the 2D QR codes through a QR reader application. The reader decodes the information from the net and reports the data available with the QR code. In 2012, [29] created a unique obstacle detection system that is fully incorporated with textile structures for detecting impediments while moving in the environment. The system integrates the sensors and actuators into the dress of the PVI for

FIGURE 4.2
One- and two-dimensional barcodes.

pathfinding and obstacle detection. [25] has utilized analog-digital code (the largest storage capacity) for the storage of information about the products has been developed. This system effectively converts the text information stored inside the AD container to speech using text-to-speech. Chhajed et al. (2014) considered various issues in scanning a barcode detection using smartphones or mobile phones [30]. The problems are poor quality of images due to low resolution of mobile cameras, noise, non-illumination and distortion of the camera, etc. The proposed method even uses uneven illumination problems and objects of different sizes.

[31] has introduced a smartphone-based application for guiding a PVI by directing them to reach an object or location in an unknown environment. This system has partially observable Markov decision-making process (POMDP) to track the application state and outputs.

Though various studies have been proposed, there are certain issues that need to be addressed.

- Scanning of barcode images in complex scenarios (folded, round or cylindrical barcodes)
- Non-ideal situations such as low lighting, dim light or in dark conditions
- Scanning the barcode in upside-down position (titled position at 180°)
- Scanning barcode on the go (during movement)
- Failure notification or alert message to retake/recapture the barcode on partial or no barcode found during the capture process
- Multilingual voice over the text specified in the product database

4.3 Proposed System

In this chapter, a real-time barcode recognition system for VI/PVI people has been proposed using a smartphone application. The application, by default, captures an image of the barcode shown by the VI, and the acquired barcode image is preprocessed for enhancement. Then the module implemented in the application screens the barcode and provides necessary information about the product that the VI has chosen in terms of voice using a text-to-voice converter. The goal of this proposed system is to observe and acknowledge the barcode to extract the merchandise information. The designed application consists of seven modules:

1. Barcode capture
2. Barcode detection
3. Barcode preprocessing
4. User feedback system
5. Decoding of barcode image
6. Fetching product specification
7. Text-to-Voice conversion

4.3.1 Barcode Capture

The proposed system uses the in-built camera of the smartphone with a photography engine to capture the barcode image shown by the VI/PVI. The captured image has been forwarded to barcode pre-processing for better detection of the product specification (Figure 4.3).

4.3.2 Barcode Detection

The Hough line transformation technique [32], available with OpenCV library, is used to recognize and discover the correct position of the captured barcode image, as shown in Figure 4.4. Once the barcode has been completely scanned, the decoding of the barcode image process will be carried out. If the barcode is incompletely found or partially found, then the user feedback system is initiated to recapture the image.

4.3.3 Barcode Pre-processing

The quality of the proposed system mainly depends on the barcode scanning process. In general, a barcode scanner with a smartphone counts pixels to recognize the width and placement of a vertical bar in a barcode sign. Therefore, there is a lack of resolution that interferes with or degrades the system's performance. For the proposed system, a minimum of 200 dpi is required to process with the barcode recognition. Higher resolution may have a higher recognition task. For instance, a 1D barcode requires at least three pixels per shortest bar and a minor gap in the symbol, as shown in Figure 4.5. In comparison, the 2D barcodes require at least five pixels.

To deal with unreadable and damaged barcodes, the proposed system uses three image enhancement techniques via pre-processing methods.

Dilation – As seen in Figure 4.6, dilation refers to appending pixels to the limits of a barcode object, whereas erosion removes the pixels from the boundaries of objects.

Binarization – transforms scanned image to binary image (black and white), resulting in zero and ones, making it easier to distinguish among image boundaries as shown in Figure 4.7.

Despeckle – analyzes the pixels' density in a given barcode area to check if the pixel is blurred or there is a noise from the acquired barcode, as shown in Figure 4.8.

Sometimes there may be a possibility that the barcodes are scanned in a titled (varying angle) manner. To rectify the image for better recognition, the proposed system uses a deskewing mechanism, in which the angle is determined to rotate the acquired barcode to 90°, i.e., to straighten it out. A barcode's distinct lines can be wider or narrower than typical barcodes. So the deskewing may worsen the recognition technique by reducing image quality on straightening. Figure 4.9 shows the scanned barcode before and after the deskew process.

4.3.4 Scan Distance

The scanning distance is significant for 1D barcodes because it can affect barcode recognition by causing the device or system to scan unnecessary pixels. As a result, for optimal decoding speed and accuracy, the suggested system employs a scan distance of five pixels or more. Normally, the higher the scan distance, the faster the recognition and accuracy will be. Furthermore, the viewing angle of the barcode reader has an effect on the needed distance and reader type. Thus, high-resolution smartphone cameras are used at one foot more than the reading angle to work even better in low-light situations.

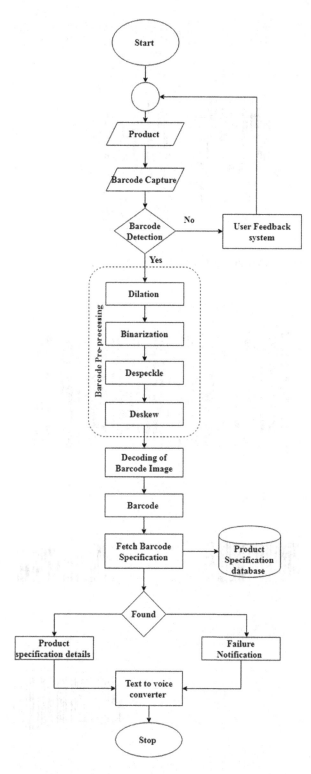

FIGURE 4.3
Transformation of barcode image from the product to voice message.

FIGURE 4.4
Hough line transformed image in a scanned barcode.

FIGURE 4.5
Barcode module width in a scanned image.

FIGURE 4.6
Scanned barcode before and after dilation.

FIGURE 4.7
Scanned barcode before and after binarization.

FIGURE 4.8
Scanned barcode before and after despeckle process.

FIGURE 4.9
Scanned barcode before and after deskew process.

For effective preprocessing, the damaged or crashed barcodes are even rectified while scanned by the PVI and VI persons to enhance the identification rates.

4.3.5 User Feedback System

Feedback is important in this system for guiding the user to capture the image. The feedback system will guide the user to recapture the cropped image if it is not recognized by the application. For the products like bottles or cans, it is harder to detect the barcodes. Mostly the vibration or beep tones facilitated with the smartphones are used for easy communication that instructs the user about the scanned image. If a partial image is detected or a barcode is damaged, instructions are given by voice to the user to rotate the merchandise in an exact direction. The text-to-speech process delivers the product information via audio to the user.

4.3.6 Decoding of Barcode Image

The ZBar library [33] is used to create a decoding engine that converts a barcode picture to the contents of the barcode. It has the capability to scan the barcodes quickly, and it supports barcodes in multiple orientations. ZBar consists of a core cross-platform barcode detector. Unlike ZXing [34], ZBar is easy to integrate and identifies the barcode quickly. It is available as a true Android library and hooks into the smartphone camera Application Programming Interface (API) and a surface view without much hassle. Further, it decodes a barcode in nano or micro seconds. Thus, the proposed system uses the ZBar to decode the barcode so that the product information can be fetched from the database at an earlier time.

4.3.7 Fetching Product Specification

To fetch the product information from the database, the barcode number generated in the decoding process is used as the search key. The query language of the proposed system is

TABLE 4.1

Sample of the Product Specification Data Stored in the Database

Barcode	Produce Name	Brand	Size	Flavor	Weight	Price	Category
8885823598512	Cosmetics	Axe	Multiple	Chocolate	100gms	196.50	Mens fashion
8856475824521	Hair Gel	Parachute	Multiple	Hibiscus	85gms	246.75	Haircare
8852142624252	Pocky	Gilco	Regular	Chocolate	47gms	15.00	Snack

implemented using SQLite. Table 4.1 shows a sample quote of the data stored in the database. For the reader's convenience, the data in other languages are translated into English.

If indeed the search is successful, the product details will be the result of the following stage. If the search fails, the VI/PVI persons will receive a voice indication about the failure.

4.3.8 Text-to-Speech Using Google TTS

The product specification details fetched from the datastore will be converted to voice using Google text-to-speech (TTS). Then the data will be delivered through the headphones/loudspeaker on the smartphone.

The resulting words are transformed into vocal using TTS technology. Google has integrated a text-to-speech API for converting text to speech for the purpose of voice recognition. Thus, the VI can use the proposed system (mobile application) to scan the barcode. The smartphone camera captures the barcode and other processes have been carried out, as said earlier. Finally, the product specification details are fetched, and the relevant textual data is converted to an audio output file in Tamil and English language according to the barcode specification database, and the same has been delivered through the VI headphone/loudspeaker of the smartphone.

4.4 Experimental Result and Discussion

In this section, the experimentation has been carried out with various categories of items to assess the effectiveness of the proposed method. Comparisons have also been made with state-of-the-art assistive hassle-free shopping methods/mobile-based barcode readers for performance analysis. The proposed system has the ability to work in various situational platforms, such as collecting the barcode in imperfect conditions, low lighting conditions, capturing and reading the barcode images on the movements, and helps the VI/PVI to depend on themselves. Other technologies such as screen readers, flatbed scanners, e-book readers, and embossers suffer from slow computational speeds, inefficient accuracy, or cumbersome usability. The proposed system has been analyzed with various barcode data captured in different non-ideal situations through a smartphone, and the dataset collection is shown in Table 4.2. The efficiency of the proposed system is evaluated using the familiar metric – recognition rate using equation 1 [35]. As the other state-of-the-art techniques have no standard benchmark dataset for online shopping, the dataset collected here has been used for the benchmarking process.

$$RR = \frac{\left(\# \text{ of tested} - \# \text{ of failure}\right)}{\# \text{ of tested}} \tag{4.1}$$

TABLE 4.2

Dataset Collection

S.No	Categories	Captured Images	Utilized Images
1	Dairy & Bakery	345	320
2	Atta, Flour Sooji	265	244
3	Snacks and Branded Foods	198	187
4	Shampoos & Conditioners	203	193
5	Beverages	145	139
6	Staples	398	386
7	Personal care	246	132
8	Home Care	197	180
9	Baby Care	124	111
10	Kitchenware	166	142
11	Pooja Items	85	79
	Total	2372	2113

4.4.1 Dataset Collection

The barcode data scanning was performed in two grocery stores in Madurai, Tamil Nadu, a southern state of India, from March to June 2021. A total of 2,113 barcodes have been scanned on non-ideal situations from various products, as shown in Table 4.2. The product has bilingual information in both English and Tamil, some product specifications are entirely in Tamil, and certain products are in both Tamil and English. Sample capture of barcode images is also shown in Figure 4.10.

The performance of the proposed system has been compared with state-of-the-art techniques such as Lock et al. [31], Chhajed [30], Kasthuri et al. [36], Jethjarurach and Limpiyakorn [24]. The proposed system achieves a recognition rate of 97.92 percent. It can be noticed from Table 4.3 that the proposed system has a higher recognition rate than the other four state-of-the-art systems. On comparing the performances, the pooja items have shown less recognition rate as the items are very small and this makes it more complex to scan the items than the other category of items by the VI/PVI. In addition to all these processes, the proposed system has the capability to scan the barcode image even in complex scenarios, as shown in Figure 4.11. The extracted barcodes are highly segmented and shown in a green color rectangle surrounding the barcodes. This proves the efficiency of the proposed system that can be helpful and adaptable to VI persons.

4.5 Conclusion and Future Work

In this chapter, a real-time barcode detection for the welfare of both VI and PVI is presented. In order to build a complete model, a smartphone-based mobile application has been designed for this proposed system. The smartphone application captures the barcode of the product shown by the visually impaired/partially visually impaired people through the smartphone camera available. The scan distance and visualization of the barcode are properly identified through the Hough transformation, and the intimation has been sent

FIGURE 4.10
Sample barcode capture on collected dataset.

TABLE 4.3

Performance Evaluation of the Proposed System with the State-of-the-Art System in Terms of Recognition Rate

S.no	Categories	Proposed System	Lock et al. [31]	Chhajed [30]	Kasthuri et al. [36]	Jethjarurach & Limpiyakorn [24]
1	Dairy & Bakery	98.6	97.23	97.36	95.21	94.32
2	Atta, Flour Sooji	97.35	96.34	98.12	95.68	93.25
3	Snacks and Branded Foods	98.66	95.21	96.44	97.21	94.65
4	Shampoos & Conditioners	98.12	96.75	97.45	95.63	94.23
5	Beverages	97.32	97.12	96.44	96.37	95.11
6	Staples	98.11	96.24	95.39	96.87	94.36
7	Personal care	98.91	96.21	96.75	95.87	94.11
8	Home Care	97.99	97.11	95.27	95.77	93.54
9	Baby Care	98.54	96.31	96.33	96.84	96.11
10	Kitchenware	97.33	97.21	97.24	96.54	95.33
11	Pooja Items	96.21	95.33	95.98	95.11	94.12
	Average RR	97.92	96.46	96.62	96.1	94.47

to the VI on partial or no barcode captured during the process. The captured barcodes are well preprocessed using dilation, erosion, binarization, deskew and despeckle processes for the highly accurate detection of barcodes. Then, the extracted image of the barcodes is decoded using ZBar and the barcode data is used to fetch the product details from the product specification database. Finally, Google text-to-speech (TTS) is used to deliver the voice message through headphones/loudspeakers of the smartphone. Also, to assess the performance of the proposed system, experimentation has been carried out with various categories of items (2,113 items from a grocery store) and the performance is compared with the state-of-the-art approaches in terms of recognition rate.

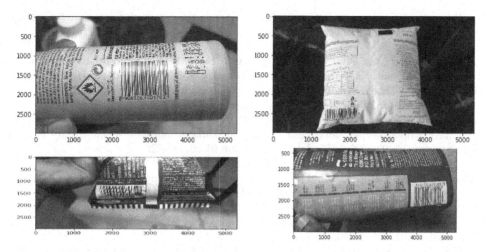

FIGURE 4.11
Barcode scanned in complex scenarios.

References

1. A. Gordois, H. Cutler, L. Pezzullo, et al. (2012). An estimation of the worldwide economic and health burden of visual impairment. *Global Public Health, 7*(5), 465–481.
2. G.A. Stevens, R.A. White, S.R. Flaxman et al. (2013). Global prevalence of vision impairment and blindness: magnitude and temporal trends, 1990–2010. *Ophthalmology, 120*(12), 2377–2384.
3. M. Zetterberg (2016). Age-related eye disease and gender. *Maturitas*, Vol. 83, 19–26.
4. D. Bouchard, and S. Tetreault (2000). The motor development of sighted children and children with moderate low vision aged 8–13. *Journal of Visual Impairment & Blindness, 94*(9), 564–573.
5. J. Graw (2009). Genetics of crystallins: cataract and beyond. *Experimental Eye Research, 88*(2), 173–189.
6. M. Manikandakumar (2020). Smart Cataract Detector: An Image Processing-Based Mobile Application for Cataract Detection. In *Computational Intelligence and Soft Computing Applications in Healthcare Management Science* (pp. 246–262). IGI Global.
7. J.T.F. Lau, V. Lee, D. Fan, M. Lau, and J. Michon (2002). Knowledge about cataract, glaucoma, and age related macular degeneration in the Hong Kong Chinese population. *British Journal of Ophthalmology*, Vol. 86, No. 10, 1080–1084.
8. P. Mitchell, G. Liew, B. Gopinath, and T.Y. Wong (2018). Age-related macular degeneration. *The Lancet, 392*(10153), 1147–1159.
9. S. Das, and C. Malathy (2018, April). "Survey on diagnosis of diseases from retinal images," *Journal of Physics: Conference Series*, Vol. 1000, No. 1, p. 012053. IOP Publishing.
10. B. Thylefors, and A.D. Negrel (1994). The global impact of glaucoma. *Bulletin of the World Health Organization, 72*(3), 323.
11. W. Elmannai, and K. Elleithy (2017). Sensor-based assistive devices for visually-impaired people: current status, challenges, and future directions. *Sensors, 17*(3), 565.
12. M. Hu, Y. Chen, G. Zhai, Z. Gao, and L. Fan (2019). An overview of assistive devices for blind and visually impaired people. *International Journal of Robotics and Automation, 34*(5), 580–598.
13. S. Sankaran, R.R. Mahaif, G. Harendra, K.N. Rao, U.S. Krishna, P.R. Murugan, and V. Govindaraj (2020, July). A Survey Report on the Emerging Technologies on Assistive Device for Visually Challenged People for Analyzing Traffic Rules. In *2020 International Conference on Communication and Signal Processing (ICCSP)* (pp. 0582–0587). IEEE.
14. R. Tapu, B. Mocanu, and T. Zaharia (2020). Wearable assistive devices for visually impaired: A state of the art survey. *Pattern Recognition Letters, 137*, 37–52.
15. K. Manjari, M. Verma, and G. Singal (2020). A survey on assistive technology for visually impaired. *Internet of Things, 11*, 100188.

16. R. Proenca, A. Guerra, and P. Campos (2013). A gestural recognition interface for intelligent wheelchair users. *International Journal of Sociotechnology and Knowledge Development (IJSKD)*, 5(2), 63–81.
17. K.M. Varpe, and M.L. Dhore (2020). A Review of Scene Text Recognition Systems for Blind. *International Journal of Advanced Science and Technology*, Vol. 29, No. 9 Special Issue, 2316–2323.
18. G. Laput, and C. Harrison (2019, May). Sensing fine-grained hand activity with smartwatches. In *Proceedings of the 2019 CHI Conference on Human Factors in Computing Systems* (pp. 1–13).
19. C. Mejia, K. Ciarlante, and K. Chheda (2021). A wearable technology solution and research agenda for housekeeper safety and health. *International Journal of Contemporary Hospitality Management*, 33(10), 3223–3255. doi:10.1108/ijchm-01-2021-0102.
20. G. Liu, X. Ding, C. Yu, L. Gao, X. Chi, and Y. Shi (2019, May). "I Bought This for Me to Look More Ordinary" A Study of Blind People Doing Online Shopping. In *Proceedings of the 2019 CHI Conference on Human Factors in Computing Systems* (pp. 1–11).
21. C.W. Yuan, B.V. Hanrahan, S. Lee, M.B. Rosson, and J.M. Carroll (2019). Constructing a holistic view of shopping with people with visual impairment: a participatory design approach. *Universal Access in the Information Society*, 18(1), 127–140.
22. M.F. Saaid, I. Ismail, and M.Z.H. Noor (2009, March). Radio frequency identification walking stick (RFIWS): A device for the blind. In *2009 5th International Colloquium on Signal Processing & Its Applications* (pp. 250–253). IEEE.
23. Y. Ebrahim, W. Abdelsalam, M. Ahmed, and S.C. Chau (2005, January). Proposing a hybrid tag-camera-based identification and navigation aid for the visually impaired. In *Second IEEE Consumer Communications and Networking Conference, 2005. CCNC. 2005* (pp. 172–177). IEEE.
24. N. Jethjarurach, and Y. Limpiyakorn (2014, May). Mobile product barcode reader for Thai blinds. In *2014 International Conference on Information Science & Applications (ICISA)* (pp. 1–4). IEEE.
25. S. W. Kim, J. K. Lee, B. S. Ryu, and C. W. Lee (2008, January). Implementation of the embedded system for visually-impaired people. In *4th IEEE International Symposium on Electronic Design, Test and Applications (delta 2008)* (pp. 466–469). IEEE.
26. M.A. Almuzaini, and M. Abdullah-Al-Wadud (2019, May). Medication Identification Aid for Visually Impaired Patients. In *2019 2nd International Conference on Computer Applications & Information Security (ICCAIS)* (pp. 1–6). IEEE.
27. R. Ghrist (2008). Barcodes: the persistent topology of data. *Bulletin of the American Mathematical Society*, 45(1), 61–75.
28. H.S. Al-Khalifa (2008, July). Utilizing QR code and mobile phones for blinds and visually impaired people. In *International Conference on Computers for Handicapped Persons* (pp. 1065–1069). Springer, Berlin, Heidelberg.
29. S.K. Bahadir, V. Koncar, and F. Kalaoglu (2012). Wearable obstacle detection system fully integrated to textile structures for visually impaired people. *Sensors and Actuators A: Physical*, 179, 297–311.
30. M.G. Chhajed (2014). Barcode detection from barcode images captured by mobile phones: an android application. *International Journal on Recent and Innovation Trends in Computing and Communication*, 2(4), 814–819.
31. J.C. Lock, A.G. Tramontano, S. Ghidoni, and N. Bellotto (2019, September). ActiVis: Mobile Object Detection and Active Guidance for People with Visual Impairments. In *International Conference on Image Analysis and Processing* (pp. 649–660). Springer, Cham.
32. I. Culjak, D. Abram, T. Pribanic, H. Dzapo, and M. Cifrek (2012, May). A brief introduction to Open CV. In *2012 proceedings of the 35th international convention MIPRO* (pp. 1725–1730). IEEE.
33. ZBar, http://zbar.sourceforge.net
34. ZXing, http://www.code.google.com/p/zxing/
35. M. Dahi, and N. Semary (2015). "Primitive Printed Arabic Optical Character Recognition Using Statistical Features," *IEEE Seventh International Conference on Intelligent Computing and Information Systems (ICICIS)*, 81–85.
36. R. Kasthuri, B. Nivetha, S. Shabana, M. Veluchamy, and S. Sivakumar (2017, March). Smart device for visually impaired people. In *2017 Third International Conference on Science Technology Engineering & Management (ICONSTEM)* (pp. 54–59). IEEE.

5

Traffic Density and On-road Moving Object Detection Management, Using Video Processing

Ankit Shrivastava and S. Poonkuntran

School of Computing Science and Engineering, VIT Bhopal University, Madhya Pradesh, India

CONTENTS

5.1 Introduction

As the population of modern cities is growing day by day, the movements of vehicles are increasing, which leads to the problem of congestion. Traffic congestion has caused many important difficulties as well as challenges in the larger and more populated cities. Increased traffic has resulted in increased waiting times and wastage. Because of these congestion problems, people lose time, miss opportunities and are frustrated.

Many issues have arisen as a consequence of the rise in automobile traffic. For example, consider traffic accidents, traffic congestion, air pollution caused by vehicles, etc. Congestion on the roads has proven to be a formidable obstacle. Upgrades in the city's pre-existing transportation infrastructure, such as additional sidewalks as well as longer roads, were generally acknowledged as failing to alleviate traffic congestion. Numerous academics have focused on intelligent transportation systems (ITS), such as predicting congestion using activity monitoring at junctions. A growing dependence on traffic monitoring is

DOI: 10.1201/9781003206736-5

needed for improved identification of cars across a wider region in order to better comprehend traffic flow. For practical purposes like traffic analysis as well as security, it is crucial to be able to recognize cars automatically from surveillance data.

A significant practical use of computer vision, such as traffic analysis or security, is vehicle identification as well as tracking in video surveillance data. Video cameras are a low-cost way to keep an eye on things. Trying to manually revise all of the data they produce is almost impossible. The best approach is to use video analysis algorithms that do not need much human input. Video surveillance systems concentrate on the modeling of the backdrop, the categorization of moving vehicles, as well as the tracking of these vehicles. Video sensors and high-performance video processing equipment are becoming more widely available, offering new possibilities for addressing video comprehension issues like vehicle tracking and target categorization. When we talk about vehicle tracking as well as grading, we are talking about a system that separates moving vehicles into various categories.

Sensors are used extensively in traffic management and information systems to estimate traffic characteristics. A wide range of traffic data, including vehicle categories, lane changes, and more, may be calculated in addition to the number of vehicles. To get started, all that is needed is the camera's calibration settings as well as the traffic flow direction. Monitoring visitors patterns and figuring out accident records from real-time video facts have two common matters:

- Video data have to be portioned and transformed into vehicles.
- The conduct of these motors is observed for instantaneous decision-making.

5.1.1 Problem Definition

The most current traffic monitoring techniques include manual or electromagnetic loop vehicle counting. In addition to being prohibitively costly, this method has the drawback of just counting.

The current image processing technique uses an activist differentiation technique, which does not allow the complete extraction of vehicle shapes. In the feature-based tracking method, the extracted image is blurred, the contour active method is very hard to implement and tracking based on the region takes a long time. In this adaptive background project, subtraction detection technique and Otsu's method are used to overcome all these problems.

5.2 Literature Survey

In this research work presented, video surveillance cameras are installed along the roads and road interactions for collecting traffic data. The information is then analyzed to get traffic parameters like road traffic density. This research work presents a comfortable and stylish approach for estimating the road traffic density during daytime using image processing and computer vision algorithms. The video data collected is first weakened into frames, which are then preprocessed during a series of steps [1]. This research work presents it is easier to get a job now because technological advancements, modernization, and instructional development have increased the number of job opportunities.

Different applications as well as administrative frameworks have a comparable impact on metro urban residents' daily lives, even in regions with a small number of residents. By getting automated frameworks in every possible division, a significant part of urban areas is now transforming into smart urban areas. Due to increased red-light deferrals, this structure reduces the likelihood of crowded traffic situations. Here, the structure has been designed with the goal of clearing traffic in accordance with vehicle density [2]. Researchers present the technique described here as designed to help users avoid or alleviate the effects of traffic congestion. The main emphasis of the device would be the picture recorded by the camera. To determine the density, the recorded picture would be compared to a pre-loaded server picture. The intersections are triggered by traffic movements based on density. As a consequence, there is less time spent waiting as well as more traffic moves smoothly. The system would run on its own, depending on data from density pictures sent to the server by the spot [6]. This research work presents outlined a method for calculating traffic density in order to improve traffic flow. In the suggested approach, vehicles entering or leaving a lane may be counted by looking at the exit area. The presented approach performed well in low-density traffic as well as in instances when the traffic did not cross paths during the experimental assessment.

To improve vehicle recognition accuracy, we want to experiment with additional detection techniques that include machine learning in the near future [10]. Researchers focus on India's traffic light management which has long been a source of frustration. Overpopulation, carelessness, and unscientific techniques have combined to lead to poor traffic signal administration. To enhance traffic conditions, new technologies and a fresh strategy are required. The density-based automated traffic control system may be a solution to this issue. By examining the traffic density at the signal, the model may adjust the traffic signals. The Raspberry Pi board is connected to a camera as well as traffic lights. With the Raspberry Pi, the footage can be analyzed in real time, and the traffic signals may be changed as needed. This may help traffic flow smoother and save time. This is an upgrade over traditional traffic signal timer functioning [14]. In this research work, at its inception, traffic system management made a commitment to improving the existing traffic control situation as well as ensuring that everything went off without a hitch. For vehicle tracking, the reliability of video detection technology has opened up new frontiers. Although this kind of traffic signaling is more organized, researchers are still a long way from seeing the broad use of video detection traffic management. There is a hefty up-front expense associated with RFID equipping as well as upkeep since each location must be individually programmed. It verifies great accuracy, unlike any other method, and we are sure of its effectiveness as well as practicality [17]. This present work discusses how roadways are congested due to increased automobile traffic. In particular, the rise in urban populations, increasing urbanization, as well as the country's economic growth has all contributed to an increase in the demand for automobiles. A fundamental difficulty in visitors' engineering functions and the wise transportation gadget is incident detection. An automatic visitor monitoring machine that can substitute or minimize human site visitors monitoring is the aim of the scheme described here. With the introduced mathematical morphological method, a transferring car may also be detected and tracked in instant time with information delivered at the control management center. The effects of the experiments indicate that the technique furnished may also be used with a number of site visitor scenarios, weather circumstances, and lighting. The present work discusses a real-time site visitor monitoring and management gadget that is reliable and semi-automated. To identify modifications in traffic patterns, and estimate traffic in recorded video, a morphological method of image processing is used. The semi-automated device receives visitors' video from the motorway as input and then analyses each picture in turn by setting a threshold value. The system's output is an alert message derived from the traffic pattern's characteristics. Depending on the complete quantity of automobiles over or below the specified threshold, two messages are shown to manage traffic, namely "Traffic" and "Normal." In addition to reducing or replacing human traffic controller duties, this automation process has the edge over others because of its high degree of precision. It is a less expensive alternative to the senor for traffic analysis since it is not as widely utilized as it formerly was. These Morphological edge detectors are also used in the experimental study, which compares them to other edge detector types, such as Sobel and Roberts. They are also compared to Prewitt's edge detector and the conventional bottom subtraction technique. The technique described here is effective as well as produces excellent results since it can monitor traffic in a variety of weather and lighting situations [20].

5.3 Technical Concepts

5.3.1 Image Processing

Using image processing, you may enhance or extract valuable information from a picture after it has been converted to digital form as well as perform operations on it. If the input is an image, such as a video frame or a picture, and the output is a picture or attributes connected with that image, then it is a kind of signal distribution. When it comes to image processing, most systems use two-dimensional outputs and use well-established signal processing techniques.

It is now one of the fastest-growing technologies, having uses across the board for a business. Engineers as well as computer programmers alike turn to image processing as a vital area of study.

It enhances an image or extracts information or features from an image. Computerized routines for extracting information (e.g., pattern recognition, classification) from remote sensing images to obtain categories of information about specific characteristics (Figure 5.1).

Image processing essentially includes the following three steps.

- Use an optical scanner or digital camera to capture the picture and upload it to your computer.
- Research and modify the picture to detect models that are not human eyes, such as satellite photos, by analyzing as well as compressing the information.
- Output is the last stage in which the image analysis result may be changed into a new picture or report.

Digital picture manipulation is made possible via the use of computer-based processing methods. There are gaps in the raw data from the imaging sensors on the satellite system. In order to correct these issues and ensure that the material is authentic, it will be subjected to several stages of treatment. Preprocessing, augmentation as well as presentation, and extraction of information are the three main stages that all kinds of digital data must go through.

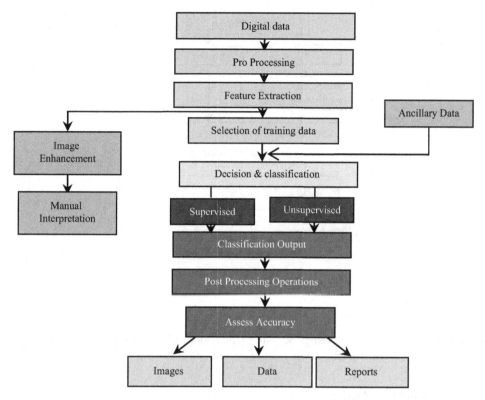

FIGURE 5.1
Flow diagram of traffic data analysis.

5.3.2 Architecture Design

Figure 5.2 gives an overview of the detection of the moving vehicle in a video sequence. The method makes use of an already existing video. The frame of reference is the initial framework. We will use the following images as our starting point. They are put up against each other, and the context is removed. An automobile in the entrance frame will be kept. Many methods follow the discovered vehicle, including adaptive background as well as blob analysis.

Assume that the initial frame of the adaptive background removal method is the backdrop of the video clips under consideration. Figure 5.2 illustrates the algorithm's design. The video clip is played back and transformed into frames using the background removal method, as shown in Figure 5.3. To begin, the difference between the first and second frames is computed, i.e., FR1 minus FR1 plus j. Step 2 compares the changes, then step 3 eliminates pixels with the same values throughout the frame difference. There are four

FIGURE 5.2
Overview of vehicle detection and counting system.

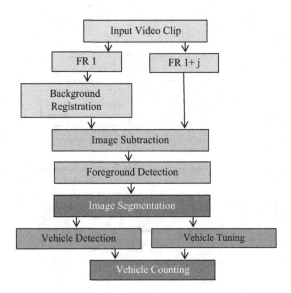

FIGURE 5.3
Vehicle detection and counting.

steps in this process: picture post-processing (the fourth phase), vehicle identification (the fifth phase), vehicle tuning (the sixth phase), as well as counting (the final phase).

5.3.3 Background Registration

Using a learning background modeling method to identify salient areas of a given video clip is a common detection strategy. To do this, you must first remove all of the images from the scene's backdrop. First, the backdrop is assumed, and then a different image is created, with the threshold used to decide which picture will be in front of the background. Pixel groups that move in unison are known as vehicles, and they may be either lighter regions on darker backgrounds or darker regions against lighter ones.

Vehicle detection is made more difficult when the vehicle's color is the same as the backdrop, or if a portion of it has aged with the background. As a result, there are an excessive number of incorrect cars.

5.3.4 Color Identification

Vehicle color identification uses a defamed color pattern. In the defamed color model, the color is identified using the intensity of the threshold images. To separate chromatic and lighting components, color spaces may be used. However, doing so might lead to an unreliable pattern, particularly for vehicles with extreme brightness or dark conversions. We are brought back to the RGB color space by the need to preserve the intensity parts while also saving money on IT. We must think of a color scheme that distinguishes between the defamed elements and brightness, just as we must detect shadows in motion.

5.3.5 Data Flow Diagram for Detection Modules

Figure 5.4 shows the video clip as the input and playback video is converted into frames. The converted frame is different from the foreground frame and the background frame. The step of processing the frame differences performed on the image using the image

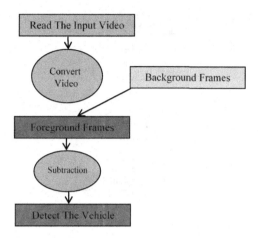

FIGURE 5.4
Data flow diagram for detection of vehicle.

subtraction and the historical past is removed, retaining solely the vehicle in the faced view and the count of the detected vehicle. Many applications, such as video surveillance, traffic monitoring, and vehicle tracking, may benefit from mobile vehicle identification and video analysis. Frame variation and adoptive background subtraction are popular motion segmentation methods.

5.3.6 Data Flow Diagram for Tracking Modules

Detecting the vehicle can convert to a grayscale image, and the grayscale image is converted to a binary image (Figure 5.5).

The binary image is the segmentation of the regions of the vehicle of interest, after the detection of vehicle regions containing unknown objects must be detected.

The extraction of the appropriate characteristics is the vehicle and then the extraction of vehicle tracking. The tracking vehicle may occur due to the sudden movement of the vehicle, changes in vehicle appearance and scene patterns. Monitoring is usually done in the context of higher-level purposes requiring the location and/or shape of the vehicle in each

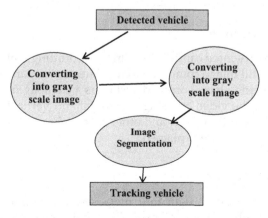

FIGURE 5.5
Data flow diagram for tracking vehicles.

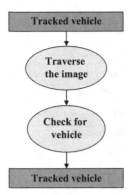

FIGURE 5.6
Vehicle counting.

frame. Identified vehicle and select the unidentified pixel on each side of the object and change the color of the pixel. Each multiple mobile vehicle present in the frame will be followed successfully.

5.3.7 Data Flow Diagram for Counting Modules

The binary image mask followed types the entered image for the count. The image is scanned from pinnacle to bottom to become aware of the presence of an automobile. When a car is detected, the picture is scanned from top to bottom. Only two factors are kept track of: the count (which maintains tabs on exactly how many cars there are) and the count register (which keeps track of which vehicles are registered).

You should always verify whether a new vehicle is already registered in the stamp before treating it as a whole new one and increasing the count. If it is not, then it is regarded as part of an already existing vehicle as well as its existence will be overlooked. The idea is utilized in all of the images, and the vehicle's final count is stored in the variable. It is possible to attain a respectable level of counting precision. In some instances, two cars are combined and regarded as a single unit owing to obstructions (Figure 5.6).

5.4 Proposed Methodology

This chapter shows the presented car detection and car counting method in the videos. The detection of cars in the still image compared to the detection of cars or objects in the video contains difficulty. The video-based object detection is a challenging task for the researchers to detect an object in the video by using horizontal and vertical edge-based methods. The presented method steps, flow chart and other descriptions are shown below. First, the steps of the presented method are explained.

5.4.1 Proposed Method Steps

The given approach is concerned with four purposes. The initial feature is to consider video recording and partition it into frames. The second characteristic is to use simple methods such as raster variations and pick out the image recorded in the background.

Following that, the backdrop is removed, leaving only the foreground cars, and the last feature aids in counting the detected automobiles. The structure of the proposed method has three parts. The first is the preprocessing task, the application algorithm, and the last post-processing task. In the preprocessing task, play the video from the dataset, split into the frame (Figure 5.7).

First step – Select the video from the dataset, and read the video file. For selecting the image form dataset employing a MATLAB function that is ui get file. Ui get file is the predefined function in MATLAB for selecting the dataset of the image. Also select the directory of the image with the help cd command.

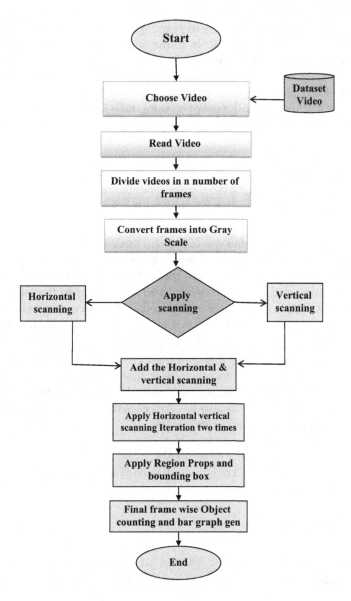

FIGURE 5.7
Proposed method.

Initialize Array M Array [] to zeros.

Declare two universal values m, n to incorporate the row or column elements of a video frame for i = 1 to Frame no. in steps of 1 with an interval of four frames, respectively. Read as well as shop each frame of the video clip.

M Array [should be used to contain the video images].

Increase the value of the variable k, which keeps track of the total number of M Array frames.

Second step – After reading the video, divide this video into the n number of frames for further processing of the video because that is not possible to directly proceed with videos. That is why first convert into the n frames, then each frame works as a single image. Apply algorithm to detect the object.

Third step – The method's purpose is to plan an effective highway number system. The initial issue with a video clip is frame segregation. After that, each photograph is displayed as an impartial RGB image that is then converted to a grayscale image. After the fine video conversion to n frames number. First, convert all frames to RGB grayscale with the MATLAB free help set rgb2gray command. During the first three steps, the preprocessing task is complete. After that, apply an object identification algorithm.

```
For i = 1:k
Change the RGB photos purchased in step 3 to grayscale. Kept them in
three-dimensional array T [m, n, 1].
End
```

Initialize array variable to play the video and store two matrix values that are rows and columns of the video frame. Read each image from video clips, then store the images in the table and then increment the table position to store the next frame; this process continues until the last frame is read and stored in the table. The image is converted to RGB as a gray image and stored in the three-dimensional array where m and n are the row and column value is given to the particular number.

Fourth step – To detect the object in the video frames, apply horizontal and vertical scanning in each frame. In this step, first create the background of the video frames. Because we want to detect the object, first create the platform for identifying the objects. For this apply image processing basic concept like converting into the unit8 and multiple with 0.1 and 0.9. After creating the background, apply the horizontal scanning of all the frames one by one.

```
//Input: d is a specific Video Frame
//Output: An image with Foreground Vehicles is saved in 'c'
//Initialize c array to zeros
for i = two to m-1
for j = two to n-1
if
difference between the pixel values of d array with bg or b array is
much less than 20 store value zero for that pixel in array c
else1
```

```
store pixel cost
end if
end for
end for
Convert the values in c array and observe median filter.
Show the output images c, d.
```

The image of the foreground of the vehicle is stored in the table variable c is initialized to zero if d is the specific video frame. The distinction in pixel value of d array with the background picture b & array < 20 save the zero cost for the difference between the pixels in array c; otherwise, store the pixel value. Convert the stored value c array and apply the median filter to remove the noise.

Fifth step – After completing the horizontal scanning of all the frames, apply vertical scanning in all the frames of the targeted video.

Sixth step – After completing the above steps 4 and 5, in the next step combining all the horizontal and vertical scanned frames of the video. For combining the horizontal and vertical scan frames apply the logical end operator to the horizontal and vertical scanned frames.

```
Scan = horz &vert; // Syntax of combining the frames
```

Seventh step – Apply the multiple iterations to the combined output frames, on the basis of the hit and trial method. After the second and third iterations, get a better output video.

Eighth step – When got the desired output of the video, the final task is computing the total number of cars in the video frames; for this, the major important operation is performed. First, apply the region proposed and second bounding box. With the help of these two MATLAB operations, generate the sharp and accurate output of the result.

Ninth step – Apply counting with the help of the above step. In the previous step, generate the rectangular frames, and use this frame to calculate and count the total number of cars. At least one step shows the total number of counting of the objects or cars in the targeted video. To represent the complete quantity of automobiles in the video and visitors' density on the road, use the bar graphs; this graph shows the per frame traffic intensity. For a better understanding of the proposed technique, also draw a flow chart. The flowchart of the proposed method is discussed below.

```
Algorithm count ()
//Initialize count=0
Counted register buffer counter=0
Step 1: Traverse automobile picture to discover a car from pinnacle
to bottom.
Step 2: If the vehicle has met, then check for registration in
counter.
Step 3: Count and register the vehicle in the counter, tagged with
the new meter if it is not authorized.
Step 4: Repeat steps 2 through 4 until the bushing is not completed.
```

The binary image mask followed1 varieties the input image for the count. To check for the presence of a vehicle, this picture is scanned from pinnacle to bottom. There are two factors kept track of: the remember, which keeps track of the number of vehicles, and counts issue count reg, which keeps track of the data of the registered vehicle. When a new vehicle is seen, check to see if it is already registered in the stamp; if it is not, it is presumed to be a new vehicle, as well as the number is increased; however, it is treated as part of a shared vehicle, as well as its presence is ignored.

5.5 Simulation and Result

The implementation of vehicle detection and counting on traffic video using image processing with simulation software is as follows:

First result – For selecting the object of the targeted video, first convert videos into the frames and then apply background clearing. Figure 5.8 shows the background cleared output frame that is frame number 72. In this clearly see that the whole background part will be black apart from cars.

Second – In a similar way next task is to convert the car or objects in the white color. Figure 5.9 shows the background and object separately with zero and one. This background contains pixel value zero in the eight-bit color image, and the object contains pixel value of one in the eight-bit grayscale image.

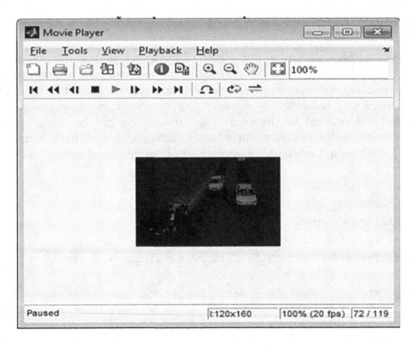

FIGURE 5.8
The background clearing.

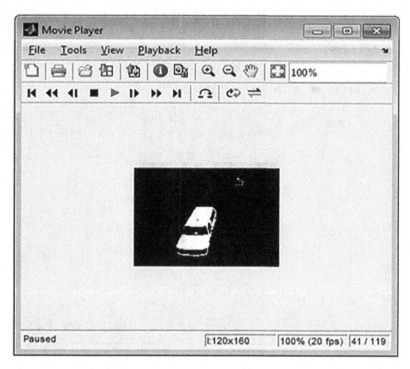

FIGURE 5.9
Frame output of object detection.

After separating both background and object, now detect the motion in the next step of showing the results figure.

Figures 5.10 and 5.11 shows the vertical scanning of the objects. Vertical and horizontal scanning is used to detect the motion of the objects.

After completing the horizontal and vertical scanning of the image, Figure 5.12 shows the addition of the horizontal scanning of the frames.

After completing horizontal and vertical scanning of the video frames and adding the result, apply the iteration process to the outcome. Again, calculate the horizontal and vertical scanning of the output, then add the horizontal and vertical scanning. Apply this same phenomenon at least two or three times for proper detection of object or target (Figure 5.13).

After applying the iteration to achieve better object detection, next is the region proposed. Apply region props and detect the proper object in the video frames. Figures 5.13 and 5.14 shows the final outcome of the proposed method. In the final result, calculate the traffic intensity of the frames.

After showing the above preprocess of the results that are shown in Figures 5.15 and 5.16. The final output result of the video is shown in Figure 5.9, which is the combination of all the above results. In the above result, clearly see that all the objects or cars are shown in the yellow box and frame by frame cars counting is shown in Figures 5.15 and 5.16.

Figures 5.15 and 5.16 show the frame-wise traffic intensity of the cars on the road. With the help of this, Figures 5.15 and 5.16 analyzed the traffic density on the road, and this graph is useful for traffic checking.

Figure 5.15 shows only the traffic on the road in terms of total number of cars.

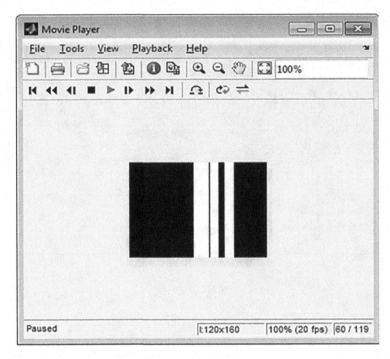

FIGURE 5.10
The vertical scanning of the video.

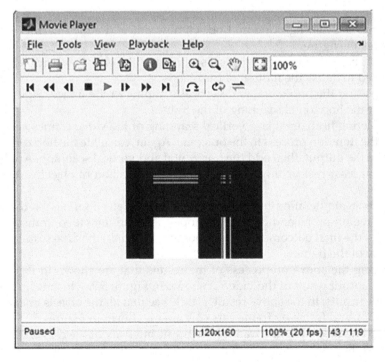

FIGURE 5.11
The horizontal scanning of the video.

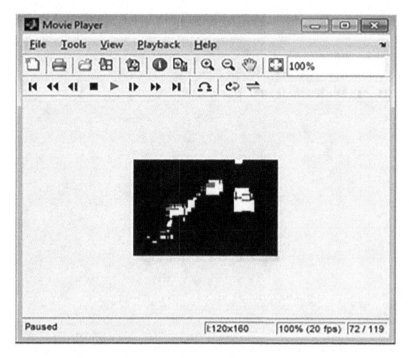

FIGURE 5.12
The addition of horizontal and vertical scanning.

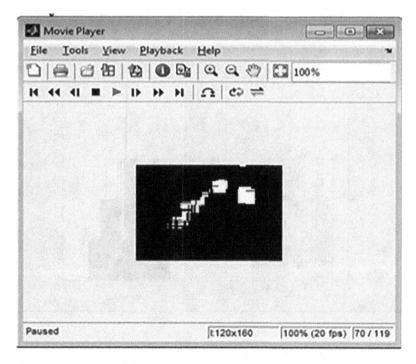

FIGURE 5.13
Iteration of the video frames.

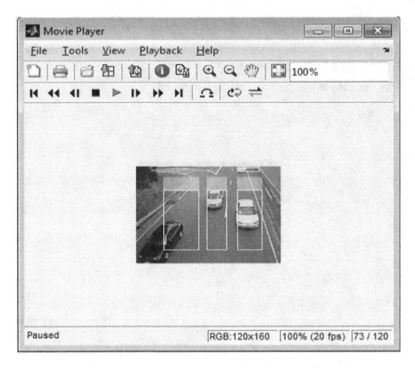

FIGURE 5.14
The final result of the proposed method.

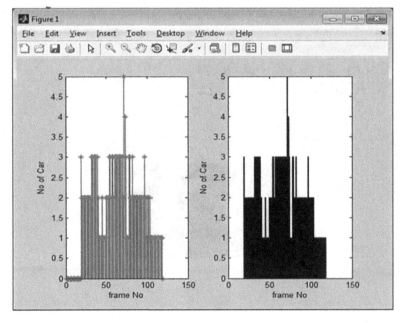

FIGURE 5.15
Graphical view of traffic intensity.

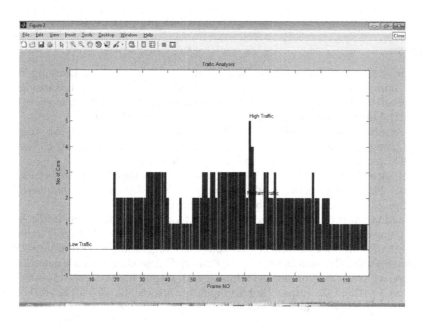

FIGURE 5.16
Traffic on the road, and low, high, and average traffic intensity.

Figure 5.16 shows the three different conditions:

1. **Low traffic** – If no. of cars is between zero and one, the traffic is low on the road.
2. **Average traffic** – If no. of cars is between one and three, the traffic is average on the road.
3. **High Traffic** – If no. of cars > 3 then the traffic is high on the road.

5.6 Conclusion

In order to effectively identify and count moving cars on the road, a system has been created. To target specific cars even when full occlusions, as well as confusing postures, exist, the system uses fundamental area know-how about car instructions combined with time area statistical measure. Background bulk is successfully rejected as a result. Although the detection of the vehicle is 100 percent, the testing findings show that the efficiency of counting cars was 96 percent, which is attributable to partial occlusions.

As the video frame size and number of identified cars increase, the method becomes more computationally demanding. When it comes to highway traffic, there's no denying that things like trees cast a shadow, yet owing to occlusions, two cars may become one and be handled as if they were one.

References

1. Suseendran, G., et al. "Incremental Multi-Feature Tensor Subspace Learning Based Smart Traffic Control System and Traffic Density Calculation Using Image Processing." *2021 2nd International Conference on Computation, Automation and Knowledge Management (ICCAKM). IEEE,* 2021. https://ieeexplore.ieee.org/abstract/document/9357743

2. Gupta, Ayushi, et al. "Real-Time Video Monitoring of Vehicular Traffic and Adaptive Signal Change Using Raspberry Pi." *2020 IEEE Students Conference on Engineering & Systems (SCES). IEEE,* 2020. https://ieeexplore.ieee.org/abstract/document/9236731

3. Agarwal, Anant, et al. "Efficient Traffic Density Estimation Using Convolutional Neural Network." *2020 6th International Conference on Signal Processing and Communication (ICSC). IEEE,* 2020. https://ieeexplore.ieee.org/abstract/document/9182718

4. Jonnalagadda, Murthy, Sashank Taduri, and Rachana Reddy. "Realtime Traffic Management System Using Object Detection based Signal Logic." *2020 IEEE Applied Imagery Pattern Recognition Workshop (AIPR). IEEE,* 2020. https://ieeexplore.ieee.org/abstract/document/9425070

5. Tasgaonkar, Pankaj P., Rahul Dev Garg, and Pradeep Kumar Garg. "Vehicle detection and traffic estimation with sensors technologies for intelligent transportation systems." *Sensing and Imaging,* vol. 21, no. 1 (2020): 1–28. https://link.springer.com/article/10.1007/s11220-020-00295-2

6. Frank, Anilloy, Yasser Salim Khamis Al Aamri, and Amer Zayegh. "IoT based smart traffic density control using image processing." *2019 4th MEC International Conference on Big Data and Smart City (ICBDSC). IEEE,* 2019. https://ieeexplore.ieee.org/abstract/document/8645568

7. Hu, Hongyu, et al. "Traffic density recognition based on image global texture feature." *International Journal of Intelligent Transportation Systems Research* vol. 17, no. 3 (2019): 171–180. https://link.springer.com/article/10.1007/s13177-019-00187-0

8. Fratama, Rizki Ramadhan, et al. "Real-Time Multiple Vehicle Counter using Background Subtraction for Traffic Monitoring System." *International Seminar on Application for Technology of Information and Communication (iSemantic). IEEE,* 2019. https://ieeexplore.ieee.org/abstract/document/8884277

9. Sentas, Ali, Seda Kul, and Ahmet Sayar. "Real-Time Traffic Rules Infringing Determination Over the Video Stream: Wrong Way and Clearway Violation Detection." *International Artificial Intelligence and Data Processing Symposium (IDAP). IEEE,* 2019. https://ieeexplore.ieee.org/abstract/document/8875889

10. Chowdhury, Md Fahim, Md Ryad Ahmed Biplob, and Jia Uddin. "Real time traffic density measurement using computer vision and dynamic traffic control." *2018 Joint 7th International Conference on Informatics, Electronics & Vision (ICIEV) and 2018 2nd International Conference on Imaging, Vision & Pattern Recognition (icIVPR). IEEE,* 2018. https://ieeexplore.ieee.org/abstract/document/8641039

11. Zhu, Jiasong, et al. "Urban traffic density estimation based on ultrahigh-resolution UAV video and deep neural network." *IEEE Journal of Selected Topics in Applied Earth Observations and Remote Sensing* vol. 11, no. 12 (2018): 4968–4981. https://ieeexplore.ieee.org/abstract/document/8536405

12. Alpatov, Boris A., Pavel V. Babayan, and Maksim D. Ershov. "Vehicle detection and counting system for real-time traffic surveillance." *2018 7th Mediterranean Conference on Embedded Computing (MECO). IEEE,* 2018. https://ieeexplore.ieee.org/abstract/document/8406017

13. Ke, Xiao, et al. "Multi-Dimensional Traffic Congestion Detection Based on Fusion of Visual Features and Convolutional Neural Network." *IEEE Transactions on Intelligent Transportation Systems* vol. 20, no. 6 (2018): 2157–2170. https://ieeexplore.ieee.org/abstract/document/8451957

14. Ashwin, S., et al. "Automatic control of road traffic using video processing." *2017 International Conference on Smart Technologies For Smart Nation IEEE,* 2017. https://ieeexplore.ieee.org/abstract/document/8358631

15. Chung, Jiyong, and Keemin Sohn. "Image-Based Learning to Measure Traffic Density Using a Deep Convolutional Neural Network." *IEEE Transactions on Intelligent Transportation Systems* vol. 19, no. 5 (2017): 1670–1675. https://ieeexplore.ieee.org/abstract/document/8011496

16. Santhosh, Kelathodi Kumaran, Debi Prosad Dogra, and Partha Pratim Roy. "Real-time moving object classification Using DPMM for road traffic management in smart cities." *2017 IEEE Region 10 Symposium (TENSYMP). IEEE,* 2017. https://ieeexplore.ieee.org/abstract/document/8070028

17. Islam, Md Rokebul, et al. "An Efficient Algorithm for Detecting Traffic Congestion and a Framework for Smart Traffic Control System." *2016 18th International Conference on Advanced Communication Technology (ICACT). IEEE,* 2016. https://ieeexplore.ieee.org/abstract/document/7423566

18. Mehboob, Fozia, Muhammad Abbas, and Richard Jiang. "Traffic event Detection from Road Surveillance Video is Based on fuzzy logic." *2016 SAI Computing Conference (SAI). IEEE,* 2016. https://ieeexplore.ieee.org/abstract/document/7555981

19. Gu, Xiao-Feng, et al. "Real-Time vehicle detection and tracking using deep neural networks." *2016 13th International Computer Conference on Wavelet Active Media Technology and Information Processing (ICCWAMTIP). IEEE,* 2016. https://ieeexplore.ieee.org/abstract/document/8079830

20. Anuradha S. G., K. Karibasappa and B. Eswar Reddy, "Morphological Change Detection System for Real Time Traffic Analysis." *International Journal of Signal Processing, Image Processing and Pattern Recognition* vol. 8, no. 8 (2015), pp. 143–150. https://ieeexplore.ieee.org/abstract/document/7449928

21. Saravanan, S. "Implementation of efficient automatic traffic surveillance using digital image processing." *2014 IEEE International Conference on Computational Intelligence and Computing Research. IEEE,* 2014. https://ieeexplore.ieee.org/abstract/document/7238419

22. S. Bhardwaj and A. Mittal, "A Survey on Various Edge Detector Techniques", *2nd International Conference on Computer, Communication, Control and Information Technology (C3IT-2012) on February 25–26, 2012, Procedia Technology,* vol. 4, (2012), pp. 220–226. https://www.sciencedirect.com/science/article/pii/S221201731200312X

23. L. Ding and A. Goshtasby, "On the Canny edge detector." *The Journal of Pattern Recognition Society PERGAMON,* vol. 34, (2001) pp. 721–725. https://www.sciencedirect.com/science/article/abs/pii/S0031320300000236

24. M. Abo-Zahhad, R. Gharieb, S. Ahmed and A. Donkol, "Edge Detection with a Preprocessing Approach." Journal of Signal and Information Processing, vol. 5, (2014), pp. 123–134. https://www.scirp.org/html/4-3400360_50479.htm

25. A. Mogelmose, Antonio Prioletti, Mohan M. Trivedi, Alberto Broggi, and Thomas B. Moeslund. "Two-stage part-based pedestrian detection." *2012 15th International IEEE Conference on Intelligent Transportation Systems.* 2012. https://ieeexplore.ieee.org/abstract/document/6338898

26. E. Stringa, "Morphological Change Detection Algorithms for Surveillance Applications." British Machine Vision Association, (2000), In BMVC, pp. 1–10. http://www.bmva.org/bmvc/2000/papers/p41.pdf

27. A. Alshennawy and A. A. Aly, "Edge detection in digital images using fuzzy logic technique." World Academy of science, engineering and technology, vol. 51, (2009), pp. 178–186. https://www.freeprojectsforall.com/wp-content/uploads/2018/10/Edge-Detection-Techniques-using-Fuzzy-Logic.pdf

28. P. A. Hajare and P. A. Tijare, "Edge Detection Techniques for Image Segmentation." International Journal of Computer Science and Applications, vol. 4, no. 1, (2011). http://www.researchpublications.org/IJCSA/issue8/2011-IJCSA-138.pdf

29. M. Nosrati, R. Karimi, M. Hariri, and K. Malekia, "Edge Detection Techniques in Processing Digital Images: Investigation of Canny Algorithm and Gabor Method." World Applied Programming, vol. 3, no. 3, (2013), pp. 116–121. http://ijses.com/wp-content/uploads/2017/11/265-IJSES-V1N10.pdf

30. Pranay Yadav. "Color image noise removal by modified adaptive threshold median filter for RVIN." In *2015 International Conference on Electronic Design, Computer Networks & Automated Verification (EDCAV),* pp. 175–180. IEEE, 2015. https://ieeexplore.ieee.org/document/7060562

31. Shachi Sharma, and Pranay Yadav. "Removal of fixed valued impulse noise by improved Trimmed Mean Median filter." *2014 IEEE International Conference on Computational Intelligence and Computing Research*, pp. 1–8. IEEE, 2014. https://ieeexplore.ieee.org/document/7238368

32. Pranay Yadav, and Parool Singh. "Color impulse noise removal by modified alpha trimmed median mean filter for FVIN." *2014 IEEE International Conference on Computational Intelligence and Computing Research*, pp. 1–8. IEEE, 2014. https://ieeexplore.ieee.org/abstract/document/7238369

33. Bharti Sharma, Sachin Kumar, Prayag Tiwari, Pranay Yadav, and Marina I. Nezhurina. "ANN based short-term traffic flow forecasting in undivided two-lane highway." *Journal of Big Data* vol. 5, no. 1 (2018): 1–16. https://journalofbigdata.springeropen.com/articles/10.1186/s40537-018-0157-0

34. J. Zhang and C. H. Chen, "Moving Objects Detection and Segmentation in Dynamic Video Backgrounds." Technologies for Homeland Security, 2007 IEEE Conference on 16–17 May 2007, pp. 64–69, E-ISBN: 1-4244-1053-5, Print ISBN:1-4244-1053-5. https://ieeexplore.ieee.org/abstract/document/4227784

35. K. Ganesan and S. Jalla, "Video Object Extraction Based on A Comparative Study of Efficient Edge Detection Techniques." International Arab Journal of Information Technology, vol. 6, no. 2, (2009) April. http://citeseerx.ist.psu.edu/viewdoc/download?doi=10.1.1.295.6793&rep=rep1&type=pdf

36. S.-C. S. Cheung and C. Kamath "Robust techniques for background subtraction in urban traffic video." Visual Communications and Image Processing, vol. 5308, no. 1, (2004), pp. 881–892. https://www.spiedigitallibrary.org/conference-proceedings-of-spie/5308/0000/Robust-techniques-for-background-subtraction-in-urban-traffic-video/10.1117/12.526886.short?SSO=1

37. E. Nadernejad, S. Sharifzadeh and H. Hassanpour, "Edge detection techniques: evaluations and comparison." Applied Mathematical Sciences, vol. 2, no. 31, (2008), pp. 1507–1520. http://www.m-hikari.com/ams/index.html

38. V. Kastrinaki, M. Zervakis and K. Kalaitzakis, "A survey of video processing techniques for traffic applications." Image and vision computing, vol. 21, no. 4, (2003), pp. 359–381. https://www.sciencedirect.com/science/article/abs/pii/S0262885603000040

6

Automated Vehicle Number Plate Recognition System, Using Convolution Long Short-Term Memory Technique

S. Srinivasan
Nehru Institute of Technology, Coimbatore, India

D. Prabha
Sri Krishna College of Engineering and Technology, Coimbatore, India

N. Mohammed Raffic
Nehru Institute of Technology, Coimbatore, India

K. Ganesh Babu
Chendhuran College of Engineering & Technology, Pudukottai, India

S. Thirumurugaveerakumar
Kumaraguru College of Technology, Coimbatore, India

K. Sangeetha
Panimalar Engineering College, Chennai, India

CONTENTS

DOI: 10.1201/9781003206736-6

6.1 Introduction

Deep learning approaches have produced good results in the computer vision area in recent years, notably for problems such as object detection, as well as identifying their class by providing a variety of deep network models [1–3]. These methods have opened the road for academics to employ strong deep learning models to build more performing algorithms and real-world systems, such as those used in license plate identification [4–7]. In smart intelligent systems, ALPR has received remarkable attention, and it is very popular because it can be applied in real-life case studies. By recognizing license plates, it is possible to identify or recognize vehicles by their unique registration numbers. The content in the number plate can be extracted using image processing techniques. The number plate content provides details such as the state and district where the motor vehicle was registered. The type of the vehicle, such as private, commercial, foreign, government or military, can be identified based on the number plate color and font color. We can get the vehicle owner's name and address from the vehicle registration number. Identifying the owner of a vehicle may be useful when rules are violated. It can also be used [5]:

- i. To find blacklisted cars
- ii. To find vehicles crossing the limits
- iii. To keep track of vehicle movements in order to combat crime
- iv. In border control systems
- v. In traffic management decisions
- vi. To understand behavior to detect anomalies

Manually inspecting such a large number of moving cars is extremely difficult, which is why the development of an ALPR system that can be fast, accurate, and can also extract car numbers from moving vehicles is essential for the development of intelligent transportation systems.

The ALPR problem may be broken down into three subtasks: Detect license plates, segment license plates, and recognize characters. The software requires seven primary algorithms [8]:

- i. From the picture, the plate is identified and isolated using the localization process
- ii. Dimensions are adjusted based on the skew of license plate
- iii. Image brightness, contrast, and saturation are adjusted to normalize the image
- iv. Each character in the plates is detected and segmented
- v. Optical character recognition is used to recognize the characters
- vi. Based on rules specific to each country, characters are checked via syntactical/geometrical analysis
- vii. The recognition value is averaged over multiple fields or images

Despite the existence of various techniques, the problem remains unsolved because of a wide range of difficulties, including [8]:

- The file resolution is poor due to a distant plate or poor-quality camera.

- Images that are smeared as a result of movement.
- Low contrast and insufficient illumination caused by reflections, overexposure, or shadows.
- The plate is obscured by an object, generally a tow bar, or dirty.
- Check for different license plates on both ends when pulled by other vehicles.
- Due to a vehicle lane shift, the camera's angle of view changes during license plate reading.
- A distinctive typeface widely used for vanity plates.
- Techniques of eluding detection.
- There is a dearth of inter-country or inter-state cooperation. Two automobiles from different countries or jurisdictions may have the same license plate number, but the plate design differs.
- The ALPR system makes many predictions for the same plate, producing a significant number of false LPs.
- Different plate designs (due to climatic conditions).

The following stages are involved in Automatic License/Number Plate Recognition (ANPR/ALPR):

Step #1: Detection and locating vehicle number plate from an input image or frame.

Step #2: Take the characters from the license plate and put them together.

Step #3: Recognize the retrieved characters using any type of optical character recognition (OCR).

Advances in deep learning, notably CNN and LSTM, have provided some helpful hints on how to deal with this issue. Here, the aim is to develop novel and powerful ALPR as per convolutional LSTM (ConvLSTM). It is used to recognize the little characters on license plates. Many real video streams were used to train and test the new approach. The experimental findings demonstrate that the proposed ALPR approach is more accurate and faster than currently available approaches, making it suitable for use in real-world settings. The paper's primary contributions are as follows: Proposing a ConvLSTM for recognizing license plates under the following circumstances: i) different colors of number plates, ii) composed in various languages, iii) different textual styles, iv) different background colors and different background images.

The main objectives of the proposed system are:

1. Detect a vehicle's number plate from a video or an image.

2. Crop the Number Plate Region.

3. Perform Preprocessing and Segmentation of the Number Plate Image.

4. Using the Dataset, create a Deep Learning Model.

5. Take the number and characters off the number plate and write them down.

The following is a breakdown of the work: Section 2 gives a quick overview of current ALPR approaches. The proposed methodology is presented in section 3, explaining system architecture. In section 4, ALPR performance and comparison with ConvLSTM are provided, and conclusions are given in the last section.

6.2 Literature Review

6.2.1 Related Work Related to License Plate Recognition Technology

In P. Prabhakar [4], recognition of license plates is proposed. This method consists of the following phases: The camera's picture is used as the input and is transformed into grayscale photos. Hough lines are identified using the Hough transform, and the quantity of linked part is cut down using segmentation of grayscale picture created by identifying edges for smoothing image, and then connected part is computed. Finally, a single character is recognized within the registration code. The goal is to show that the proposed methodology obtained high accuracy by optimizing a number of factors, resulting in a greater recognition rate than traditional methods.

Ashok Kumar Sahoo [5] proposed the automation on finding the Indian number plates. To extract all the information from the license plate, the design includes license plate segmentation and character extraction, as well as segmented character identification. The focus of this study is on Hindu arabic numerals written in Latin letters. In the Prewitt filter approach, letters are extracted from the license plate and segmented in a connected component analysis. This study uses three different classifiers: k-nearest neighbors, artificial neural network, and decision tree. It is possible to achieve recognition accuracy of up to 98.10 percent, which is very promising for future research in this area.

Xavier et al. [9] elaborates license plate detection and recognition (LPDR). Here, a hybrid approach was developed to identify candidate license plates for a particular region. This study provides a kernel density function approach combined with binary preprocessing techniques. The filtered binary value of the image is used to determine the place of the vehicle sign. The method far exceeds advanced strategies in terms of performance. The elaborated technique recognized license plates (LP) with 98.1 percent accuracy and a calculation time of 0.452 seconds.

Hendry and Chen [10] used YOLO's seven folding layers to identify one class in this study. The sliding window technique is used for detection. The purpose is to identify the license plate of Taiwan. He used the Application-Oriented License Plate (AOLP) 6-digit license plate number. Sliding windows recognize all digits, and it is recognized by one YOLO structure.

6.2.2 Deep Learning-Based Work for Recognizing License Plates

- Artificial neural network [11] – Türkyilmaz has proposed a powerful system for recognizing vehicle license plates. He proposed three feedforward ANNs to recognize characters. It receives data from the outside world and feeds it to a processor. Reach the level of results compared to the base model.

- Convolutional neural networks (CNNs) – A very popular way to recognize letters on license plates. [12] CNN is very effective and can recognize license plates under various lighting conditions and with a very short processing time. To handle multi-national car license plates of various colors and designs, Muhammad

RizwanAsif [13] offers an approach termed novel illumination invariant. Since the license plate is near the taillights, the red corona is first used to identify the car's taillights and create the area of interest. Vertical edges inside each interesting area are created using a method for retaining the edges and to improve performance. The region of the license plate is distinguished using a heuristic energy map. High-level characteristics derived from the AlexNet convolutional neural network are utilized to validate the identified areas. Extensive testing on license plates in different countries has shown that it performs well. Silva [14] has proposed "end-to-end ALPR" based on a hierarchical CNN. The main concepts are to use two paths on the same CNN to recognize vehicle and license plate areas, and then a second CNN to recognize the letters. His approach was evaluated on publicly accessible datasets of Brazilian and European license plates, and it outperformed both rival academic methods and a commercial system in terms of accuracy. Naaman Omar [15] proposed a cascade deep learning approach to build an ALP system for detecting vehicles in the northern part of Iraq. This license plate includes three regions: license plate, urban area, and national area. The proposed model uses the CNN model, and the performance was evaluated. The result shows that the technique was effective. Lixin Ma et al. [1] recommend using vehicle manufacturer identification (VMR) to reduce the need for precise identification and study of CNN systems. The liner is used depending on the data received from CNN. The question extension technique is used to train NN, to extend synthetic transformation and increase recall rate. This technique is not adaptable to the environment and can maintain excellent accuracy as light and noise conditions change. The dataset showing that this method can be significantly generalized does not show a high degree of classification accuracy according to experimental results. The average accuracy of the class is better than previous methods, demonstrating that it is good in identifying vehicles by benchmarking [16]. The proposed method is based on CNN. It is used to find self-synthesized features with 90 percent accuracy [17]. We proposed a CNN-based method with SIFT (scale-invariant feature transformation) used to find local features in images. This method effectively removes fake license plates and recognizes vehicle license plates with an accuracy of 84.3°. Qian proposed an LPR with CNN technology [18]. It is able to identify the characters of the number plate effectively, and its performance rate is 93 percent. Zhu implemented a system with CNN used to identify the vehicle from the video sequences with accuracy of 82.5 percent [19]. [20] proposed a method with CNN. This method focuses on a joint denoising and rectification approach. It achieves 93.08 percent accuracy.

- Generative adversarial networks (GANs): Gupta proposed a model designed using GAN used to identify the image from high-resolution number plates [21]. Wang proposed a model which combines GAN, CNN, LSTM and BRNN (Bi-directional Recurrent Neural Network) [22]. It is used to find the number plates from moving vehicles. It achieves 89.4 percent accuracy. Zhang proposed the Cycle GAN method for registration plate recognition [23]. They used CNN encoder used to find the number plates with different designs. The proposed method achieves 80 percent.

- Recurrent neural networks (RNNs) [24] proposed a system that combines RNN + CNN with 95 precisions. CNNs are used for feature extraction, and RNNs are used for calculations. From this, I was able to understand the combination of RNNs, and CNN got good results.

6.3 Methodology

This section introduces the ConvLSTM model for LP localization and also describes some of the related steps in processing. The first step is to convert the captured image to gray-scale format and preprocess it to remove unwanted noise. Next, numbers and letters are extracted from the preprocessed image. The associated pixels are improved, the background pixels are weakened, and super-resolution technology is used to obtain a well-segmented image. ConvLSTM, a type of RNN, is utilized to recognize objects from continuous images or videos. The Attention Layer ConvLSTM technique has been implemented to mitigate the vanishing gradient problems faced by traditional RNNs. Tests are run with different dropout rates for accuracy. A solution for identifying and recognizing license plates using deep learning models has been developed and evaluated by Keras.

6.3.1 Convolutional LSTM

ConvLSTM is a hybrid neural network used to detect objects such as sequences in images. This is a fusion of deep CNN and LSTM [25].

The ConvLSTM model accommodates a simple convoluted neural network (CNN) by maintaining convolution layers and max-polling layers while deleting the fully connected layer to retrieve the sequence of features from an input picture. The CNN release is the entry of the LSTM built to make labeling for each frame of the sequence of features.

a. [26] Learn sequence function with CNN
CNN can extract deep features from images. Pooling, convolution, normalization, and layers that are fully connected make up the CNN model. These layers are utilized in a particular order to build diverse architectures for different tasks. Local features are extracted from the input pictures using convolution layers.

Let z_j^{l-1} indicates the function of the previous layer. a_{ij}^l is the learnable kernel, b_j^l is the bias to reduce overfitting. M_j is the input map selection and f(.) is an activation function. The feature map Z_j^l is calculated using Equation (6.1),

$$Z_j^l = f\left(\sum_{i \in M_j} z_j^{l-1} \times a_{ij}^l + b_j^l\right) \tag{6.1}$$

b. Sequence labeling using LSTM
In sequential issues like handwriting recognition and language translation, the LSTM has had much success [22]. An LSTM neural network is developed on top of CNNs to learn contextual information in license numbers. Instead of using an RNN unit, an LSTM is used to avoid gradient vanishing. A memory cell and gates make up an LSTM, which is a sort of RNN unit. LSTM consists of three gates to process the inputs: i) forget gate, ii) input gate, iii) output gate. Figure 6.1 shows the internal structure of LSTM. x_t refers to current input, h_{t-1} refers to previous cell output, h_t refers to current cell output, C_t refers to current cell state.

i. Forget gate f_t is represented using Equation 6.2:

$$f_t = \sigma\left(W^{(f)}x_t + U^{(f)}h_{t-1} + b^{(f)}\right) \tag{6.2}$$

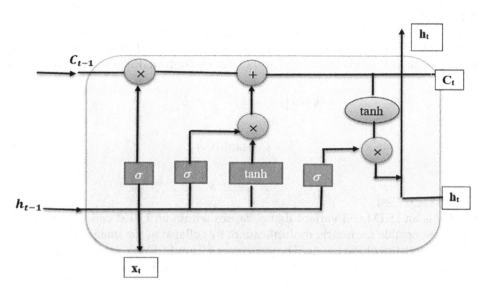

FIGURE 6.1
LSTM structure.

where $W^{(f)}$ is the weight of input x_t for forget layer, $U^{(f)}$ is hidden layer weight h_{t-1}, $b^{(f)}$ is bias function, σ is sigmoid function output the value between 0 to 1.

ii. Input gate
In this gate, two operations take place: i) adding the information and ii) updating the information.

- New information is added only when the information is important. This is achieved by two operations. Input gate i_t decides values to be updated and $\widetilde{C}_t$ creates the new candidate values as shown in Equations 6.3 and 6.4.

$$i_t = \sigma\left(W^{(i)}x_t + U^{(i)}h_{t-1} + b^{(i)}\right) \tag{6.3}$$

$$\widetilde{C}_t = \tanh\left(W^{(C)}x_t + U^{(C)}h_{t-1} + b^{(C)}\right) \tag{6.4}$$

where $W^{(i)}$ is input weight for input x_t, $U^{(i)}$ is the weight of hidden state h_{t-1} for the input gate, $b^{(f)}$ is the input gate bias function. $W^{(C)}$ is the weight of cell state, $U^{(C)}$ weight of hidden state for cell state. The tanh function converts the input between −1 to 1.

- Old state C_{t-1} is converted to new state C_t using forget gate f_t and previous cell state C_{t-1}, input gate i_t and $\widetilde{C}_t$ new candidate values as shown in Equation 6.5.

$$C_t = \left(C_{t-1} \times f_t\right) + \left(i_t \times \widetilde{C}_t\right) \tag{6.5}$$

iii. Output gate

Output gate h_t is used to decide which information has to pass to the next cell. This will act as a filter. Two operations take place: i) σ sigmoid layer decides which information has to go out, ii) Product of $\tanh(C_t)$ and O_t decides output. Refer to Equations 6.6, 6.7

$$O_t = \sigma \left(W^{(o)} x_t + U^{(o)} h_{t-1} + b^{(o)} \right) \tag{6.6}$$

$$h_t = O_t \times \tanh \left(C_t \right) \tag{6.7}$$

c. ConvLSTM [27,28]

ConvLSTM is an LSTM cell variant that collapses within an LSTM cell. The convolution process is responsible for matrix multiplication. By collapsing the image, you capture the spatial characteristics of the image. The structure of ConvLSTM is shown in Figure 6.2.

The inputs are X1,…, Xt. The cell outputs are C1,…, Ct. Hidden states are H1,…, Ht. ConvLSTM uses the past state of the input and its local neighbors to predict the future state of a specific cell in the grid. For state to state and state to state transitions, the convolution operator comes in handy. The main calculations for ConvLSTM are given, where '*' indicates the convolution operator and '∘' indicates the Hadamard product as before. Refer to Equations 6.8 to 6.12.

$$i_t = \left(W_{xi} \times X_t + W_{hi} \times H_{t-1} + W_{ci} \times C_{t-1} + b_i \right) * \sigma \tag{6.8}$$

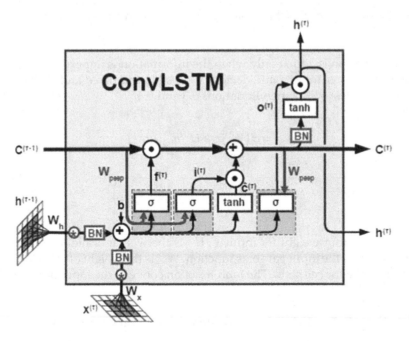

FIGURE 6.2
ConvLSTM structure.

$$f_t = \left(W_{xf} \times X_t + W_{hf} \times H_{t-1} + W_{cf} \times C_{t-1} + b_f \right) * \sigma \tag{6.9}$$

$$C_t = f_t \overset{\circ}{} C_{t-1} + i_{t-1} \overset{\circ}{} \tanh \left(W_{xc} \times X_t + W_{hc} \times H_{t-1} + b_c \right) \tag{6.10}$$

$$O_t = \sigma \left(W_{xo} \times X_t + W_{ho} \times H_{t-1} + W_{co} \overset{\circ}{} C_t + b_o \right) \tag{6.11}$$

$$H_t = O_t \overset{\circ}{} \tanh \left(C_t \right) \tag{6.12}$$

Faster motions should be captured by ConvLSTM with a bigger transition kernel, whereas slower movements should be captured by ConvLSTM with a smaller kernel.

6.4 Experiments

Steps in implementation:

1. Data preparation
2. Model design
3. Training the model
4. Model evaluation

1. Data preparation
   Alphanumeric records are used for CNN training. The ICDAR dataset [12] consists of approximately 12,000 samples. It contains a class of 10 digits, a class of 26 upper-case letters, a class of 26 lowercase letters, and many other special characters. Train a CNN model using 6,548 samples of 10 digits and 26 uppercase letters. The length of the sequence is fixed at 70 frames per video. Then extract the frames from the video according to the length of the sequence. The following code obtains the path of a single movie and takes frames from it. Some of the image samples from the dataset with the recognition results are shown in Figure 6.3
   Extraction of frames from videos:
   The object is created by recording the specified video up to the length of the sequence. Reads one frame from the provided video at a time. If the frame cannot be read, the success is 0, and the code jumps to another part. If the photo is captured correctly, resize the photo and add it to the list. The code for extraction is shown in Figure 6.4.

2. Conv-based model design
   Model input is a video frame. ConvLSTM predicts output and provides a sequence at every time step. It is used to train data using the current and previous inputs. Figure 6.5 shows the workflow of ConvLSTM, and Figure 6.6 shows the code for ConvLSTM.

3. Training the model
   Make a training model and a test model out of the data. Then compile the model with various optimizers and hyperparameter settings to get better results.

FIGURE 6.3
Image samples with results.

```
def frames_extraction(v_p):
    fms_lt = []

    vObj = cv2.VCapture(v_p)
    # Used as counter variable
    ct = 1

    while ct <= seq_l:

        success, img = vObj.read()
        if success:
            img = cv2.resize(img, (img_height, img_width))
            frames_list.append(img)
            ct += 1
        else:
            print("Defected frame")
            break
    return fms_lt
```

FIGURE 6.4
Extraction code.

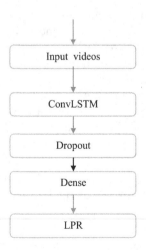

FIGURE 6.5
Workflow of ConvLSTM.

```
mod = Seq()
mod.add(ConvLSTM2D(filters = 64, kernel_size = (3, 3), return_sequences = False,
data_format = "channels_last", input_shape = (seq_len, img_height, img_width, 3)))
mod.add(Dropout(0.2))
mod.add(Flatten())
mod.add(Dense(256, activation="relu"))
mod.add(Dropout(0.3))
mod.add(Dense(6, activation = "softmax"))
```

FIGURE 6.6
Code for ConvLSTM.

4. Evaluation of model
   Evaluate the model using performance metrics. In our method, we used accuracy.

6.5 Results

6.5.1 Parameters for Evaluation

Training accounts for 80 percent while testing accounts for 20 percent. For training purposes, the video's maximum length is set to 70 seconds. To learn the model parameters of the neural network, we employ the cross-entropy-based loss function of the category in our model. Dropout rates for training with ReLU and Softmax as activation functions range from 0.2 to 0.5. Using a backpropagation method with opt optimization, train the network using a learning rate of 0.0001. After each training epoch, the network is checked for valid data. The model's parameters are listed in Table 6.1.

6.5.2 Evaluation Metrics

We used classification measures to assess the model's accuracy and efficiency. True positive (TP), true negative (TN), false positive (FP), and false negative (FN) rating scales are used to assess model effectiveness correctly and erroneously (FN). True positive rate: actual value is positive and expected value is positive; false positive rate: actual value is negative and predicted value is negative. The actual value is higher than the anticipated value; thus, it is a win-win situation. False negative rate: actual value is negative, and expected value is negative. True negative rate: actual value is negative, and predicted value is negative. The value predicted is positive. Equation 6.13 shows the relation for the calculation of accuracy.

$$Accuracy = \frac{TP + TN}{TP + TN + FP + FN} \tag{6.13}$$

6.5.3 Comparison Evaluation Fusion Model With Baseline Models

The classification accuracy of the basic model and the proposed model are compared in this table. The suggested model outperforms the other basic models, as evidenced by the results in Table 6.2.Comparison Methods

TABLE 6.1

Parameters and their Values

Parameters	Value
Input length(frames)	70
Activation	ReLU and Softmax
Loss	categorical cross entropy
Optimizer	opt

TABLE 6.2

Experimental Results for the Various Deep Learning Models

Models	Accuracy
OKM+CNN	98.10%
GAN+CNN	80.00%
CNN+RNN	95.10%
GAN+DCNN+BRNN+LSTM	89.40%
Proposed (ConvLSTM)	99.17%

Compare the proposed method with other basic models in terms of accuracy. Use the following basic model for this [10]

a. **OKM+CNN:** Used to describe the characters from the number plates.
b. **GAN+CNN:** It consists of the model of cyclic GAN for image recognition. Includes 2D warning sign recognizer along with Xception-based CNN encoder.
c. **CNN+RNN:** The combined model of CNN and RNN. CNN is used for feature extraction. RNN is used for computation.

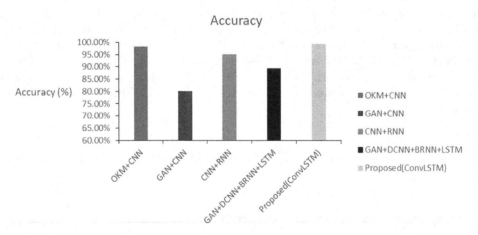

FIGURE 6.7
Experimental results of different deep learning models.

TABLE 6.3

Processing Time and Aspect Ratio

S.NO	Veh. Number	Aspect Ratio	Processing Time%
1	4PGAB5	3.837	0.262
2	5NACAD6	4.500	0.191
3	FBKPV46	4.896	0.104
4	SIGNIUDW	2.027	0.278
5	MXYKW1	3.215	0.123

d. **GAN+DCNN+BRNN+LSTM:** Use GAN, DCNN, BRNN, and LSTM for learning and sequence identification.

Table 6.2 shows the test results of various models. And those plots are visualized using diagrams and shown in Figure 6.6. The accuracy of the scoring metric is used to score the ConvLSTM model. With five approaches, our approach outperforms all models (Figure 6.7).

i. **Result of processing images**
   Table 6.3 below shows the experimental results obtained from the processed vehicle images using the processing time and aspect ratio.

6.6 Conclusion

We proved the accuracy of the license plate recognition (LPDR) system on numerous models. The suggested technique is utilized to recognize moving vehicle license plates quickly. Several approaches are used to obtain accuracy. We propose an efficient license plate recognition algorithm based on the ConvLSTM model and numerous strategies to tackle the problem of overfitting. Experiments reveal that the suggested approach outperforms all other basic models in terms of accuracy.

References

1. Lixin, Ma, and Yong Zhang, "Research on vehicle license plate recognition technology based on deep convolutional neural networks," *Microprocessors and Microsystems*, 82, April 2021, 103932.
2. W. Weihong, and T. Jiaoyang, "Research on License Plate Recognition Algorithms Based on Deep Learning in Complex Environment," IEEE Access, 8, 2020, 91661–91675. DOI:10.1109/ACCESS.2020.2994287.
3. A. Elihos, B. Balci, B. Alkan, and Y. Artan, "Deep learning based segmentation free license plate recognition using roadway surveillance camera images," ArXiv, 2019.

4. P. Prabhakar, P. Anupama and S. R. Resmi, "Automatic vehicle number plate detection and recognition," *2014 International Conference on Control, Instrumentation, Communication and Computational Technologies (ICCICCT)*, 2014, pp. 185–190, DOI: 10.1109/ICCICCT.2014.6992954.

5. Ashok Kumar Sahoo, "Automatic recognition of Indian vehicles license plates using machine learning approaches," *Materials Today: Proceedings*, October, 20, DOI:10.1016/j.matpr.2020.09.046.

6. L. Yao, Y. Zhao, J. Fan, M. Liu, J. Jiang, and Y. Wan, "Research and Application of License Plate Recognition Technology Based on Deep Learning," *Journal of Physics: Conference Series*, 1237(2), 2019. DOI:10.1088/1742-6596/1237/2/022155.

7. N. Eswar, and D. Gowri Shankar Reddy, "Morphological Operation based Vehicle Number Plate Detection," *International Journal of Engineering Research And*, 9 (2), 2020, 428–433. DOI:10.17577/ijertv9is020064.

8. https://en.wikipedia.org/wiki/Automatic_number-plate_recognition/

9. Xavier, C. Christoper, and D. Judson, "Detection of the vehicle license plate using a kernel density with default search radius algorithm filter," *Optik*, March 2020. DOI: 10.1016/j.ijleo.2020.164689.

10. Hendry and R-C. Chen, "Automatic License Plate Recognition via sliding-window darknet-YOLO deep learning," *Image and Vision Computing*, 87, 2019, 47–56.

11. Türkyilmaz, K. Kaçan, "License plate recognition system using artificial neural networks," *ETRI Journal*, 39(2), 2017, 163–172. DOI:10.4218/etrij.17.0115.0766.

12. D. Karatzas, L. Gomez-Bigorda, A. Nicolaou, S. Ghosh, A. Bagdanov, M. Iwamura, J. Matas, L. Neumann, V.R. Chandrasekhar, S. Lu, F. Shafait, S. Uchida, E. Valveny, Icdar 2015 competition on robust reading. In Document Analysis and Recognition (ICDAR), *2015 13th International Conference on*, pp. 1156–1160, August 2015. DOI: 10.1109/ICDAR.2015.7333942.

13. Asif, Muhammad Rizwan, Chun Qi, Tiexiang Wang, Muhammad Sadiq Fareed, and Syed Ali Raza, "License plate detection for multi-national vehicles: An illumination invariant approach in multi-lane environment," *Computers & Electrical Engineering*, 78, 2019, 132–147. ISSN 0045-7906, DOI: 10.1016/j.compeleceng.2019.07.012.

14. Silva, Sergio Montazzolli, and Claudio Rosito Jung, "Real-time license plate detection and recognition using deep convolutional neural networks" *Journal of Visual Communication and Image Representation*, 71, 2020, 102773, ISSN 1047-3203.

15. Omar, Naaman, Abdulkadir Sengur, and Salim Ganim Saeed Al-Ali, "Cascaded deep learning-based efficient approach for license plate detection and recognition," *Expert Systems with Applications*, 149, 2020, 113280, ISSN 0957-4174. DOI: 10.1016/j.eswa.2020.113280.

16. M. Mondal, P. Mondal, N. Saha, P. Chattopadhyay, "Automatic number plate recognition using CNN based self synthesized feature learning," *2017 IEEE Calcutta Conference, CALCON 2017 - Proceedings, 2018-January*, 378–381, 2018. DOI:10.1109/CALCON.2017.8280759.

17. X. Yang, and X. Wang, "Recognizing License Plates in Real-Time," 2019.

18. Y.G. Qian, D.F. Ma, B. Wang, J. Pan, J.M. Wang, J.H. Chen, W.J. Zhou, and J.S. Lei, "Spot evasion attacks: Adversarial examples for license plate recognition systems with convolutional neural networks," *Computers and Society*, 95, 2020, 1–14.

19. L. Zhu, S. Wang, C. Li, and Z. Yang, "License Plate Recognition in Urban Road Based on Vehicle Tracking and Result Integration," *Journal of Intelligent Systems*, 29(1), 2020, 1587–1597. DOI:10.1515/jisys-2018-0446.

20. Y. Lee, J. Lee, H. Ahn, and M. Jeon, "SNIDER: Single noisy image denoising and rectification for improving license plate recognition," *Proceedings - 2019 International Conference on Computer Vision Workshop, ICCVW 2019*, 1017–1026, 2019, DOI:10.1109/ICCVW.2019.00131.

21. M. Gupta, A. Kumar, and S. Madhvanath, "Parametric Synthesis of Text on Stylized Backgrounds using PGGANs," ArXiv, 2018. DOI: 10.48550/arXiv.1809.08488.

22. X. Wang, Z. Man, M. You, and C. Shen, "Adversarial generation of training examples: Applications to moving vehicle license plate recognition," ArXiv, 2017, 1–13.

23. L. Zhang, P. Wang, H. Li, Z. Li, C. Shen, and Y. Zhang, "A Robust Attentional Framework for License Plate Recognition in the Wild," ArXiv, 2020, 1–10. DOI:10.1109/tits.2020.3000072.

24. T.K. Cheang, Y.S. Chong, and Y.H. Tay, "Segmentation-free Vehicle License Plate Recognition using ConvNet-RNN," ArXiv, 2017.

25. B. Shi, X. Bai, and C. Yao, An end-to-end trainable neural network for image-based sequence recognition and its application to scene text recognition. *IEEE Transactions on Pattern Analysis and Machine Intelligence*, 39, 2017, 2298–2304.

26. Erdal Başaran, Zafer Cömert, and Yüksel Çelik, (2020). Convolutional neural network approach for automatic tympanic membrane detection and classification, *Biomedical Signal Processing and Control*, 56, 101734.

27. K. Sangeetha, and D. Prabha, "Sentiment analysis of student feedback using multi-head attention fusion model of word and context embedding for LSTM," Journal of Ambient Intelligence and Humanized Computing, 12(3), 2020, 4117–4126, Impact Factor: 4.594. DOI: 10.1007/s12652-020-01791-9

28. K. Sangeetha, and D. Prabha, "Understand Students Feedback Using Bi-Integrated CRF Model Based Target Extraction," *Computer Systems Science and Engineering*, 40(2), 2022, 735–747. DOI: 10.32604/csse.2022.019310.

7

Deep Learning-Based Indian Vehicle Number Plate Detection and Recognition

M. Arun Anoop
Royal College of Engineering and Technology, Akkikkavu, Thrissur, Kerala

S. Poonkuntran
School of Computing Science and Engineering, VIT Bhopal University, Madhya Pradesh, India

P. Karthikeyan
Velammal College of Engineering and Technology, Madurai, Tamil Nadu, India

CONTENTS

7.1 Introduction

ANPR is automatic number plate recognition, consisting of two core ideas: plate detection and character extraction by the OCR method. It is helpful in the area of stolen vehicles and road safety. ANPR recognizes the number plate automatically as we know its name starts with the term "automatic" by performing OCR on images to read the license plate on vehicles. The main applications are highway monitoring, parking management, neighborhood law enforcement security and identifying over speeding based on average speed calculation criteria. The main categorization in our work is AF_pMD_{lpd}, meaning it consists of three keys. Keys are authentication, foggy image prediction, ML-based and DL-based vehicle number plate detection. The primary evaluation has been done based on these category keys (Figures 7.1–7.8).

The remainder of the chapter is coordinated as follows: Section 7.2 will introduce the literature overview. Section 7.3 will introduce the experimentation results obtained by utilizing the created vehicle number plate detection (VNPD) system. Section 7.3 (a) will specify about supervised learning-based number plate forgery detection. Section 7.3 (b) will introduce foggy image prediction based on highlights removed from those images. Section 7.3 (c) will examine ML-based vehicle prediction results. Section 7.3 (d) will

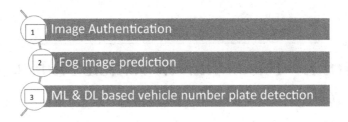

FIGURE 7.1
Flow of the chapter.

FIGURE 7.2
Object detection regular methods.

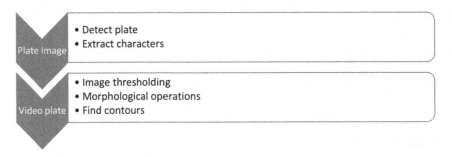

FIGURE 7.3
Some important steps.

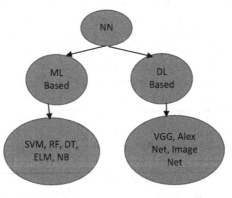

FIGURE 7.4
Neural network start-to-end flow.

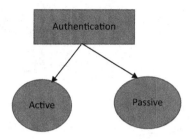

FIGURE 7.5
Authentication types.

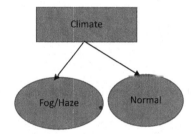

FIGURE 7.6
Climate types.

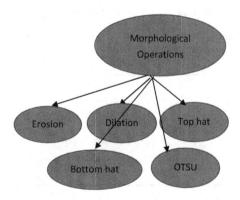

FIGURE 7.7
Morphological operations.

introduce various deep learning models and Section 7.4 examines the outcomes. Finally, the conclusions and future enhancements are discussed (Figures 7.9–7.16).

In India (Ashish Khanagwal [9]), there are mainly three types of number plates, as follows (Figures 7.17 and 7.18):

1. White-colored plates with black numbers can normally be seen on private vehicles owned for personal use.

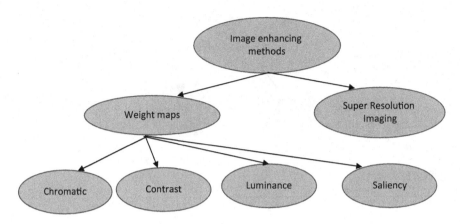

FIGURE 7.8
Image-enhancing methods.

TABLE 7.1

Performance Evaluation Based on First Proposed Method

Methodology	DT	SVM
M1	91%	<u>96%</u>

TABLE 7.2

Performance Evaluation Based on Second Proposed Method

Methodology	n-Folds (5,10)	
M2	93%	95%

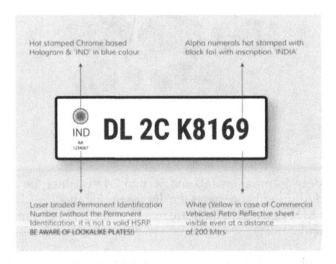

FIGURE 7.9
High-Security Registration Plate (HSRP) and color-coded fuel sticker [1].

FIGURE 7.10
Security plate features [2].

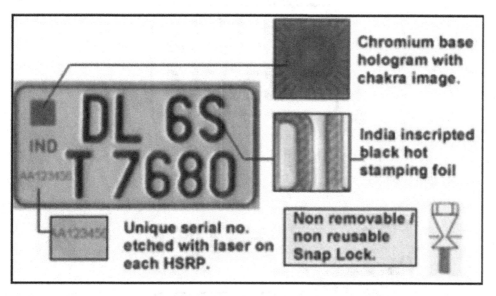

FIGURE 7.11
Delhi HSRP [3].

2. Yellow number plates with black numbers are normally used by cabs, taxis or autos.

3. Black number plates with yellow numbers can also be seen on hired cars that do not have drivers; that is, they are self-driven. These types can be seen in cars such as Zoom cars.

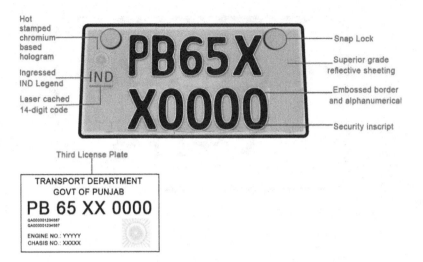

FIGURE 7.12
Punjab number plate [4].

FIGURE 7.13
Maharashtra number plate features [5].

FIGURE 7.14
Notation of Chakra and stamping foil [6].

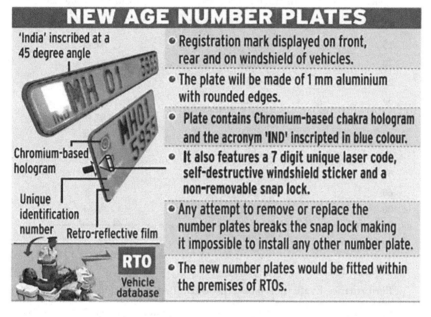

FIGURE 7.15
HSRP Features [7].

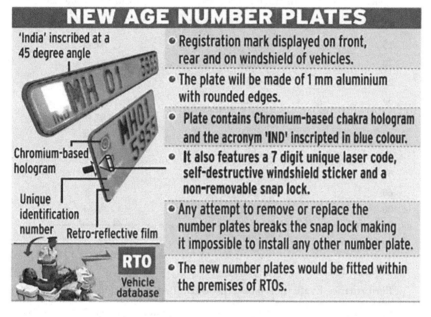

FIGURE 7.16
New Age Number Plate [8].

FIGURE 7.17
White plate.

FIGURE 7.18
Yellow plate.

TABLE 7.3

Performance Evaluation Based on Third Proposed Method

Methodology	Locating Plate and Character Extraction	
M3	Locating by bounding box	Character extraction by OCR

TABLE 7.4

Performance Evaluation

Epochs	Validation F1-score (%)	Custom F1-score (%)
10	52	47.57
15	70	64.47
20	76.34	71.64
25	83.48	80.67
30	89.14	83.80
35	91.52	87.50
40	91.82	91.67
45	91.96	93.75
50	93.45	93.63
60	95.68	93.98
80	98.07	96.64

TABLE 7.5

Performance Evaluation Based on Fourth Proposed Method

Methodology	n-Epochs (60, 80)	
M4	93.98%	**_96.64%_**
Mean (M4)	94%	

Apart from these, the following are the other important plates (Figures 7.19–7.22):

- Blue plates with white numbers are used by foreign consulates in India.
- The president's car has a red plate with national emblem and no numbers.
- Defense vehicles have black plates with white numbers and an arrow.

FIGURE 7.19
Blue plate.

FIGURE 7.20
Defense vehicle plate.

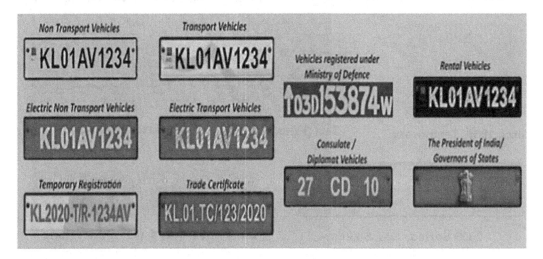

FIGURE 7.21
Different types of number plates in India [10].

On account of inappropriate number plates [12], the Motor Vehicle Act (Rule 50, 51 of MV Act, 1989) indicated in the segment below (Figures 7.23 and 7.24):

a. Registration letters and numbers will be in dim on white establishment for two-wheelers/Light Motor Vehicles (LMV) cars and black letters on yellow background for business vehicles.
b. The proportions of the number plate and the letters will be as given in the flyer for each characterization of vehicle.
c. Display of the number plate on the front and rear of the vehicle will be as shown in the flyer.
d. Fancy lettering is not permitted.
e. Other names, pictures, and articulations should not be shown.

Size of number plates in India:

1. Size of number plate for two- and three-wheelers: 200 × 100 mm
2. Size of number plate for light motor vehicles/explorer vehicles: 340 × 200 mm or 500 × 120 mm
3. Size of number plate for medium/generous business vehicles: 340 × 200 mm

Tampering/forgery/manipulations real-life survey:

Audi with counterfeit number plate seized by authorities of Andheri RTO [13].

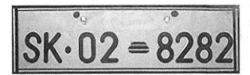

Commercially Used Vehicle, Sikkim   Commercially Used Vehicle, West Bengal

 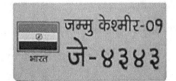 

Circa 1999, Temporary   Taxi (Hindi Script)   Taxi (Tamil Script)

2009 Series, Taxi, Sikkim   Diplomatic Corps, 68 = Switzerland

Federal Government   Government, West Bengal
Andaman and Nicobar Islands

Prime Minister and State
Governors (1)

www.worldlicenseplates.com
4/2013

Government, Sikkim

Military Vehicle

FIGURE 7.22
Other types of vehicle number plate [11].

FIGURE 7.23
Other types of vehicle number plate [26].

FIGURE 7.24
Foreign number plates [27].

Counterfeit number plate saw [14].

Lamborghini with counterfeit number plate seized [15].

Man fakes number plate in Mysore to defy traffic guidelines and make others trouble [16].

Mysore traffic police have held onto a bicycle with two diverse number plates [17].

Lady falsely utilizes Ratan Tata's vehicle number [18].

Number plate altering in Hyderabad [19].

In Hyderabad, more than 900 cases enrolled distinctly because of number plate altering in 2019 [20].

Indian entertainer Salil Acharya confronted the equivalent problem [21].

7.2 Literature Survey

Thanongsak Sirithinaphong and Kosin Chamnongthai [28] developed a recognition method for Thai car license plates, in that for car license plate extraction, they used four layers of back propagation method, and for candidate region extraction, they considered color information and shape features for the exact features extraction from the image. And finally, they achieved a performance of 84.29 and a recognition rate of 80.81 percent and they also used the sigmoid function.

Thanongsak Sirithinaphong and Kosin Chamnongthai (1999) fostered an acknowledgment strategy for vehicle tags. They fostered a programmed vehicle leaving framework and utilized Thai vehicle tags for their exploration work. Moreover, they accomplished an extraction rate of 92 percent and an acknowledgment rate of 96 percent.

Luis Salgado et al. [29] developed a LOCOMOTIVE product for car number plate detection and recognition called European ESPRIT 5184 LOCOMOTIVE project. And the main role of this product is to automatically detect license plate and its letter extraction by recognition process. Main steps processed in it are pyramid handler, region interest locating, moving object tracking, object intrusion detection, normalization, character segmentation, OCR, syntactic analyzer.

Shyang-Lih Chang et al. [30] developed an automatic license plate recognition tool for license plate recognition. For that, they used 1,065 images, of which 47 were failed during the recognition process. And they used self-organizing OCR for their work. Mainly edge and color models like edge, hue, saturation and intensity features used and after the fuzzification process utilization they were renamed as edge, hue, saturation and intensity fuzzification features. Moreover, the authors used topological sorting for locating plate correctly and additionally they used two-stage fuzzy aggregation and in that self-organizing character recognition a "Mexican hat function" is also processed for better results. In their proposed work, their product achieved success locating rate 97.9 percent, identification rate 95.6 percent and overall success rate 93.7 percent. And they also recommended using intrinsic images from scene image and number of images criteria also has to focus.

Sarthak Babbar et al. [24] proposed method for vehicle number plate detection. In that they used "Connected Component Analysis (CCA)", poly and linear kernel for support vector machine, ratio analysis, un-sharp masking and median filters for noise reduction is the main core to predict their machine learning based vehicle number plate detection.

Cheng-Hung Lin et al. [25] used new method for vehicle number plate detection. They used YOLOv2 throughout their work to detect the vehicle number plate. They used SVM machine learning algorithm combination with YOLO. They extracted features by histogram of gradients (HOG) and finally they designed YOLOv2 Darknet-19 for their vehicle number plate detection. They used a convolution neural network for improving the character recognition of blurred and obscure images.

Title	Detection Method
Recognition method for Thai car license plate (1998, 1999).	Backpropagation method. Automatic car parking system.
LOCOMOTIVE product for car number plate detection and recognition (1999).	Pyramid handler, region interest locating.
An automatic license plate recognition tool (2004).	Edge, hue, saturation and intensity fuzzification features extraction method.
Method for vehicle number plate detection (2018).	CCA, noise reduction filters and ration analysis.
A new method for vehicle number plate detection (2018).	YOLOv2 Darknet-19, CNN models, HOG and SVM.

7.3 Proposed System

In our first proposed vehicle number plate forgery detection, forged images were created using our MATLAB-based GUI module, Image Manipulation Program (GIMP) [23]. This tool is designed for researchers working in the image forgery detection field. This article is based on only copy paste and copy rotate (where θ=0, 45, 90). This image manipulation tool was created during[1] research work (Figures 7.25–7.27).

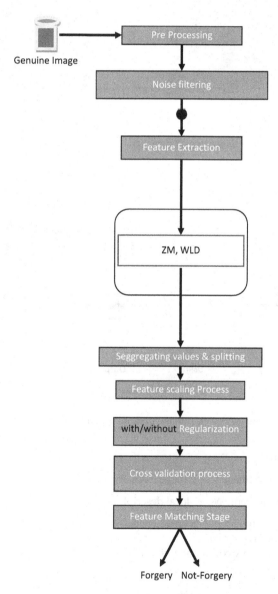

FIGURE 7.25
Proposed vehicle number plate forgery detection.

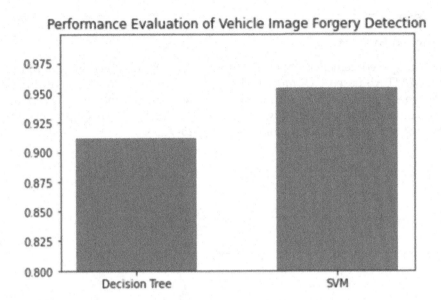

FIGURE 7.26
Performance evaluation of vehicle image forgery detection.

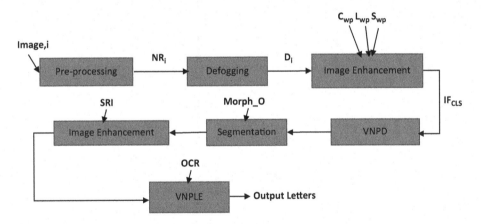

FIGURE 7.27
ML-based proposed architecture for vehicle number plate detection.

Input: Genuine image

Output: Manipulated image (copy move, copy rotate).

Features: Zernike moment, Weber local descriptors.

Condition: The operation depends on θ (where $\theta=0, 45, 90$). If θ is 0, it will be "copy move"; otherwise, it will be "copy rotate."
   If the image has x and y coordinates, it will be (Ix, Iy); notate it as Iz.

Procedure: (Fold is 10, Testing 25%)

Consider θ = 0, no rotation is applied, but a slight change happened in the particular random position. The final result will be θ (Igx1, Igy1); notate it as Imz1, where g and m are genuine and manipulated results.

If we consider θ as other input degree values, it will be manipulated based on the degree values which we performed. The final result will be θ (Igx2, Igy2); notate it as Imz2. The result varies based on θ values.

$$\left\{\begin{matrix}(Ix,Iy)\\(Imz1)\\Imz2\end{matrix}\right\} \theta = 0, 45, 90$$

Finally,

<Genuine, forged> := <(Ix,Iyie; Igx,Igy,....Igyn), (Imz1,Imz2,....Imzn)>

In our second proposed vehicle number plate forgery detection in the fog and defog situation, initially as a primer evaluation, the system has to perform the classification of two categories, mainly <fog, defog> images based on supervised machine learning approaches. For this work, fogged images were captured. For defog image collection, a hybrid form of different weight maps is processed. Weight maps fused are saliency, chromatic and luminance weight maps. For features, different Local binary pattern (LBPs) processed. Feature scaling, such as standard scaling, has been done; further, the process continues with lasso regression to avoid overfitting. Predicted accuracy is considered the average value of n-folds is 94 percent.

Input: Fog images acquisition

Output: <Fog, defog classification>

Condition: Fog images labeled 1 and defog images labeled 0.

Machine learning approach used: supervised SVM, decision trees.

n-fold cross validation: 5,10

Feature Vector (FV): <LBP's variants>

LBP variants [22] used are <Local binary pattern (LBP), Improved Local binary pattern (ILBP), Local Ternary Pattern (LTP), Mean of Constrained Least Square (Mean CLS) wp>

$$\alpha = LBP_{N,R}(x,y) = \sum_{p=0}^{N-1} s_3\left(g_p, g_c, t_1\right)2^p \tag{7.1}$$

$$ILBP_{N,R}(x,y) = \sum_{p=0}^{N-1} s\left(g_p - g_{mean}\right)2^p + s\left(g_c - g_{mean}\right)2^N \tag{7.2}$$

$$g_{mean} = \frac{1}{N+1}\left(\sum_{P=0}^{N-1} g_p + g_c\right)$$

$$LTP_{N,R}(x,y) = \sum_{p=0}^{N-1} s_3\left(g_p,g_c,t_1\right)2^p$$

$$s_3\left(g_p,g_c,t_1\right) = \begin{cases} 1 & g_p \geq g_c + t_1 \\ 0 & g_c - t_1 \leq g_p \leq g_c + t_1 \\ -1 & otherwise \end{cases} \qquad (7.3)$$

$$\text{Feature vector}(FV) = \left\{< \alpha, ILBP_{N,R}(x,y), LTP_{N,R}(x,y), \text{Mean}(CLS)wp >\right\}$$
$$= \left\{< F1, F2, F3, F4 >\right\}$$

Feature_1,2,3,4 (F1,2,3,4) is generated from the following equation,

Output image (Cwp) of {LBP}: Output image of chromatic weight map processed by LBP produced one feature, named F1.

Output image (Lwp) of {ILBP}: Output image of luminance weight map processed by LBP produced one feature, named F2.

Output image (Swp) of {LTP}: Output image of saliency weight map processed by LBP produced one feature, named F3.

And the final feature, F4, is the mean of all the three above feature algorithms. Mean (C, L, Swp).

In our third proposed vehicle number plate forgery detection, we used ML-based methods for the matching process and hybridized algorithms for feature extraction.

Input: Fog images of vehicle number plate

Output: Enhanced image and extracted number plate letters

Step 1:   Live images considered for the work. If image is Di, then {Di | i = 1, 2, 3,...n}. Where "D" is the dataset, "i" is the image, "1, 2, 3,...., n" is "n" number of images we considered for this.

Step 2:   Defogging module will convert fogged image to defog for the next process. Preprocessing step will process the image only if it is noisy, otherwise it may not apply wiener-filtering.
         In Di, the image is processed with preprocessing; finally, it results in NRi (noise removed from image).

Step 3:   In the image-enhancement phase, different weight maps are considered. Chromatic weight map, luminance weight map, saliency weight map and its fusion are also processed for the final processing. Output is notated as IF_{CLS}, which is nothing but image of fused CLS.

Step 4:   The image is processed by the vehicle number plate detection (VNPD) module, resulting in the plate location.

Step 5:   The segmentation module uses morphological operations. Morph_O applies to get the exact location of the vehicle number plate. The different types of morph_O are top-hat, bottom-hat, OTSU, erosion, and dilation.

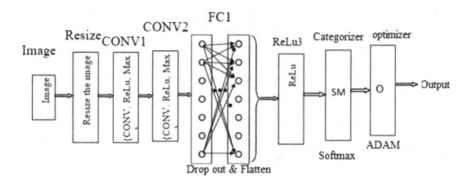

<train,val> folders.
<train folder>: 26images each 0..9A..Z
<val folder>: 6images each 0..9A..Z
Test image:car.jpeg
indian_license_plate.xml
epochs=10,15,20,25,30,35,40,45,50,60,80.
Parameters: Custom and Validation F1-score

FIGURE 7.28
Proposed deep learning-based vehicle number plate detection and recognition.

Step 6:   The image enhancement phase has to be done with super-resolution imaging for the final process to extract plate letters.

Step 7:   In vehicle number plate letter extraction (VNPLE), optical character recognition (OCR) is used to extract letters from the number plate. Finally, it results in an enhanced image and extracted number plate letters.

In our fourth proposed vehicle number plate forgery detection, we used deep learning-based methods for the matching process, and CNN was used for features extraction. Pre-trained AlexNet layers were used for the evaluation process. We utilized already existing methods to form our proposed design (Figure 7.28).

7.4 Experimentation & Results

MATLAB R2014a was used for ML-based supervised learning evaluation, and all feature extraction processes were based on MATLAB. After that, the feature matching process has done based on Google Colab. And finally, selected best among support vector machine (SVM) and decision tree (DT) in the case of methodology-1. K-fold cross validation has been processed for the primary evaluation below:

1. Forged vehicle number plate prediction
2. Fog image classification

Later ML-based vehicle number plate recognition has handled utilizing MATLAB alone. Used existing MATLAB code yet a couple of adjustments has done for getting to haze pictures. Further deep learning approaches are likewise conducted for vehicle number plate location and acknowledgment. It has done dependent on pre-trained convolutional neural network models. The means of assessment are the accompanying.

A trial was done on MATLAB R2014a (Image Processing Toolbox™) for location and acknowledgment of vehicle number plates. This identification and acknowledgment process comprises the accompanying advances.

Step 1 Figure 1: The system considers fogged images. Then the system identifies the forgery prediction of vehicle number plates. Finally, supervised ML-based approaches are used for the evaluation process.

Step 2 Figure 2: The system considers fog and defog images. Then the system identifies the classification of fog images. Performance evaluation processed and the best combination approach.

Step 3 Figure 3: The system used ML algorithm LBP for extracting features and tested live images of number plates.

Step 4 Figure 4: The system used CNN for extracting features and tested live images of number plates. In order to evaluate the experiment results, (train, val) consists of 0..9A..Z images (different categorical) were appointed. Training image set carrying 26 images of each 0..9A..Z. Validation image set carrying six images of 0..9A..Z. Different deep learning methods are considered for final accuracy prediction. Pre-trained AlexNet model used for Indian number plate detection and recognition.

References

1. HSRP and Color coded Fuel Sticker, Available: https://www.spinny.com/blog/index.php/high-security-registration-plate-full-details/
2. High security features, Available: https://autoportal.com/news/high-security-registration-plates-mandatory-from-january-1-2019-13188.html
3. Delhi HSRP, Available: https://www.bgsbuniversity.org/delhi-hsrp-number-plate-apply-online-challan-status/
4. Punjab number plate, Available: Available: https://www.hsrppunjab.com/hsrp-specifications.php
5. Maharashtra number plate features, Available: https://www.nagpurtoday.in/new-high-security-number-plates-to-come-in-2019/04232005
6. Notation of Chakra and stamping foil, Available: https://www.tesz.in/assets/guides/5ea291a64083d-img-39.jpg
7. HSRP features, Available: https://www.tesz.in/assets/guides/5ea291a63c40e-img-39.jpg
8. New age number plate, Available: http://allzhere.in/2013/06/11/high-security-number-platehsrp-process/
9. Ashish Khanagwal about Indian number plate types, Available: https://www.quora.com/What-are-the-different-types-of-number-plates-in-India
10. Different types of number plates in India, Available: mvd.kerala.gov.in
11. Other types of vehicle number plate, Available: http://www.worldlicenseplates.com/world/AS_INDI.html.
12. Improper number plate, Available: https://www.htp.gov.in/Annexure-I.pdf

13. Audi with fake number plate seized by officials of Andheri RTO, Available: https://www.team-bhp.com/forum/attachments/indian-car-scene/1574612d1 478673901-beware-fake-registration-numbers-imported-cars-proof-pg-3-psx_20161109_121143.jpg.

14. Fake number plate noticed, Available: https://www.cartoq.com/politicians-son-caught-in-range-rover-with-fake-registration-plate-victim-or-cheat/

15. Lamborghini with fake number plate seized, Available: https://gaadiwaadi.com/lamborghini-with-fake-registration-number-seized-in-bengaluru/

16. Man fakes number plate in Mysore to break traffic rules and make others trouble, Available: https://www.cartoq.com/fake-number-plate-man-arrested/

17. Mysore traffic police have seized a bike with two different number plates, Available: https://starofmysore.com/cops-seize-bike-with-different-number-plates/

18. Woman fraudulently uses Ratan Tata's car number, Available: https://www.india.com/viral/mumbai-woman-fraudulently-uses-ratan-tatas-car-number-case-exposed-after-challan-sent-to-his-office-4309770/

19. Number plate tampering in Hyderabad, Available: https://www.cartoq.com/tampered-registration-plates/

20. In Hyderabad over 900 cases registered only in the case of number plate tampering in the year 2019. Available: https://timesofindia.indiatimes.com/city/hyderabad/over-900-cases-booked-for-tampering-vehicle-number-plates-in-just-2-days/articleshow/70199188.cms

21. Indian actor Salil Acharya faced same problem, Available: https://www.latestly.com/auto/indian-actor-salil-acharya-bmws-same-number-plate-registered-on-total-three-cars-these-images-leave-twitterati-confused-2228527.html

22. Arun Anoop, M., Poonkuntran, S. LPG: a novel approach for medical forgery detection in image transmission. *Journal of Ambient Intelligence and Humanized Computing* 12, 4925–4941 (2021). https://doi.org/10.1007/s12652-020-01932-0

23. Arun Anoop M, Poonkuntran S, "Lora Approach for Image Forgery Detection and Localization in Digital Images," CnR's International Journal of Social & Scientific Research, India, Vol. 4, Issue (III) ISSN: 2454-3187, January 2019.

24. Sarthak Babbar, Saommya Kesarwani, Navroz Dewan, Kartik Shangle and Sanjeev Patel, "A New Approach For Vehicle Number Plate Detection," *Proceedings of 2018 Eleventh International Conference on Contemporary Computing (IC3), 2–4 August, 2018,* Noida, India

25. Cheng-Hung Lin, Yong-Sin Lin, and Wei-Chen Liu, "An Efficient License Plate Recognition System Using Convolution Neural Networks," *Proceedings of IEEE International Conference on Applied System Innovation 2018.*

26. Muhammad Ali Raza, Chun Qi, Muhammad Rizwan Asif and Muhammad Armoghan Khan, "An Adaptive Approach for Multi-National Vehicle License Plate Recognition Using Multi-Level Deep Features and Foreground Polarity Detection Model," MDPI, Appl. Sci. 2020, 10, 2165. doi:10.3390/app10062165.

27. Foreign number plates, Available: https://www.fontshop.com/content/license-plate-alphabets

28. Sirithinaphong, Thanongsak and Kosin Chamnongthai. "Extraction of car license plate using motor vehicle regulation and character pattern recognition." *IEEE. APCCAS 1998. 1998 IEEE Asia-Pacific Conference on Circuits and Systems. Microelectronics and Integrating Systems. Proceedings (Cat. No.98EX242)* (1998): 559–562.

29. L. Salgado, J. M. Menendez, E. Rendon and N. Garcia, "Automatic car plate detection and recognition through intelligent vision engineering," *Proceedings IEEE 33rd Annual 1999 International Carnahan Conference on Security Technology (Cat. No.99CH36303),* 1999, pp. 71–76. doi: 10.1109/CCST.1999.797895.

30. Shyang-Lih Chang, Li-Shien Chen, Yun-Chung Chung and Sei-Wan Chen, "Automatic license plate recognition," *IEEE Transactions on Intelligent Transportation Systems,* vol. 5, no. 1, pp. 42–53, March 2004. DOI: 10.1109/TITS.2004.825086.

8

Smart Diabetes System Using CNN in Health Data Analytics

P. Ravikumaran

Fatima Michael College of Engineering & Technology, Madurai, Tamil Nadu, India

K. Vimala Devi

Vellore Institute of Technology, Vellore, India

K. Valarmathi

P.S.R Engineering College, Sivakasi, India

CONTENTS

DOI: 10.1201/9781003206736-8

8.1 Introduction

The ever-increasing levels of chronic diseases but the lack of ability to prepare and obtain critical information from various health-related information is now a significant problem. There are several important reasons for accepting the renaming of healthcare components to aid in the provision of evidence-based medicine (EBM). Diabetes is the most common chronic disease, with about 8.5 percent of the planet's population enduring it; worldwide, more than 400 million people have to fight diabetes [1]. Particularly compelling is the fact that diabetes is considered a global epidemic; therefore, it is necessary to develop prevention strategies and treatments for chronic diabetes. The main distribution of this material is summarized as follows:

- Considering the different forms of data analytics in public health data.
- The need to update the quality of healthcare to test the unique features of big data analytics tools and forums.
- Demonstrating the different levels of data analysis and its key components.
- To articulate the challenges of open research facing the healthcare area and potential plans and directions.

8.1.1 What Is Big Data?

Enormous datasets are exceptionally vast and cutting-edge data, so the PC framework cannot handle it [1]. The expression "huge information" generally alludes to the utilization of prescient measurements, client conduct insights, or other progressed information examination strategies that concentrate value from information, and are once in a while in a particular informational collection size. Huge data challenges incorporate photography, distant area, information examination, appearance, sharing, trade, photographs, polls, advancement and data security [1]. The principal reference point has five levels: volume, velocity, and variety, along with the recently added veracity and value, as shown in Figure 8.1. Large data is a developing innovation; hence, information datasets and documents are extremely challenging to oversee the executive frameworks for general purposes.

8.1.2 Analytics in Big Data

Because of understanding impedance in huge information, we should rethink and ponder its ease of use, PC intricacy, and calculations. The furthest down the line, enormous PC innovation should add to more prominent PC-focused, novel and exceptionally effective PC frameworks for overseeing and examining a lot of data and keeping up with value-driven applications across all areas. New features in big data processing, such as insufficient samples, open and uncertain data relationships, and unbalanced distribution of value density, not only provide great opportunities, but also pose grand challenges, to studying the computability of big data and the development of new computing paradigms. The following steps clearly describe how we can extract and use information according to our needs:

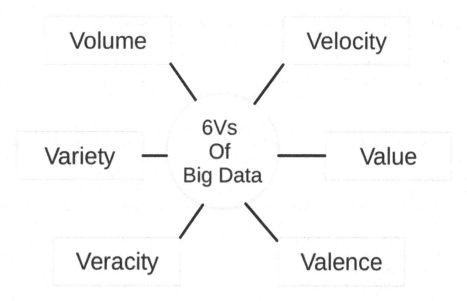

FIGURE 8.1
The 6Vs of big data.

- First, there must be precise requirements and differentiation under special, social, or financial conditions.

- Alternatively, in order to work with large amounts of data, we will need to identify and determine the composition of the kernel or part of the information to be managed. Finding kernel information and properties is no small feat because it is special to a particular domain.

- Next, a management show from above should be adopted. Although the upcoming route may allow us to skip a few career problems, it will eventually not be integrated into meaningful plans.

- Eventually, all the results will be combined to find a beneficial solution instead of different phase outcomes.

8.1.3 Healthcare – Big Data Analytics

Huge information data in healthcare includes dealing with the huge and irksome arrangements of data and making it effectively available. It is seriously engaging when it is well used in medical care investigation since it already gives us thought regarding contaminations, which licenses experts to act proactively. Using quantifiable computations can significantly improve higher, more grounded, improved, and best medications.

Here is the progression of information procurement and the executives in the medical services industry:

- Initially, data is accessible from an assortment of sources like cell phones, sensors, patients, centers, examiners, medical services suppliers, and associations in plain view that make up a tremendous measure of clinical data.

- The mix of data from different sources will be remarkable, from electronic health records (EHRs), medical imaging (MI), prescription reports (PR), genomic arrangement (GA), clinical records (CR), and medical diagnosis (MD).

- They are then haphazardly positioned between the return medical server (MS), the clinical site database (CDB), and other clinical data records (CDR) for reassessment.

- The capacity of the force structures is to store, fix, investigate, process and recover a lot of data to support the local area.

Thus, it will be useful to recognize the manifestations, ailment, and inclusion; likewise, observing will bring about diagnosing the illness in the beginning phases and settling on informed choices. The instruments are valuable for basic ways of diagnosing and treating patients critically. The reason here is to ask about the significance of enormous information data just as the different advances joined into AI techniques in medical services. Insightful techniques work to recognize the objectives of diagnosing, treating, separating, and treating all patients needing medical assistance. The arranging of extensive information in medical care should be improved as the powerful systems showed by the medical care classifications should be additionally evolved. Since patients are the last beneficiaries of names in the medical care framework, they need to settle on monetary benefit and informed choices to secure themselves. As at no other time, these difficulties should be attended to in the medical services climate ahead of time. Figure 8.2 shows the model statistic analysis system.

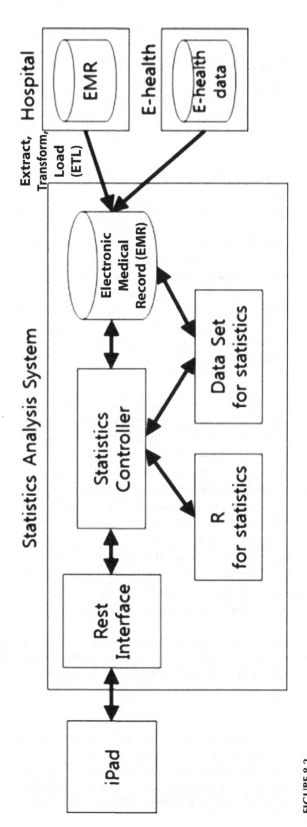

FIGURE 8.2
Statistics analysis system.

8.1.3.1 Challenges

"The principal difficulties of big data analytics in medical services are taking care of, looking, sharing and breaking down medical services data. Illustrative decisions are often made through massive data analytics. The chain of data in medical services actually requires progression since the skilled techniques presented by medical services sections must be redesigned" [2].

8.1.3.2 Developing Complexity of Healthcare Information

"Keeping up with medical care data is barely convincing since medical services data is kept in medical services records, they're enormous in nature and desire a lot of room, in advance, on the off chance that related data may provide insights with examination to support empirical data. Crafted by technocrats in the medical services field is to ensure change among trade and development, which makes way for huge information in medical care information outline works. Public accomplices are as yet having a discussion roughly the main thankful to digitize getting records, and heaps of associations have somehow at any point to make the change from paper to electronic health records (EHRs). Accordingly, data analysts and different agents consistently make some outrageous memories in getting the information they need to run pivotal business smart applications. In EHR systems, the organization's data examiners in a manner of speaking feature more profound inquiries on clinical consideration that was once in the past not present" [2].

1. The motivation behind medical care associations is to expand a conceivably dependable arrangement, which requires the distinguishing proof of perilous entanglements, while others perceive that patients who should be restored might be accessible to assist with withdrawal.

2. It is currently normal practice to change an individual who picks proof-based medical care. New EHR developments require further consolidation to find experts and clinical offices. The most recent improvements have appeared as cutting-edge specialized devices like sensors, versatile applications and recording contraptions.

3. With the advances of big data, medical care has been well established since the last decade and is having an effect in the most ideal manner. This big data idea is utilized to break down and endure data at various rates to give the normal outcomes.

4. Enormous data analytics in medical care finds exceptional procedures to further develop care, saving lives for minimal price.

To make the significance of medical care, big data analytics is basic to medical services to give quick attention to patients to advance prompt help from drug glut. This advance makes extraordinary and exceptionally available assets for patients.

8.1.4 Big Data Framework

Medical services data is a bunch of genomic data, clinical data and conduct data. Coordinated EHR, unstructured EHR and helpful pictures are creative consequences of clinical data. In changing all medical services data, some of the optional instruments and strategies are incorporated, such as organized and unstructured EHR data, innate data, therapeutic imaging data, hereditary information and other information (the investigation of illness transmission and behavioral).

TABLE 8.1

Platforms and Description

Tools/Platforms	Description
Hadoop Distributed File System	HDFS is a passed on record system that handles colossal instructive assortments running on thing gear. It helps scale a single Apache Hadoop bundle to hundreds (and surprisingly a colossal number of) hubs [8].
MapReduce	MapReduce is a model inside the Hadoop structure that is used to move to gigantic dataset aside in the Hadoop File System (HDFS).It is a middle part, essential to the working of the Hadoop structure.[8]
PIG & PIG Latin	PIG is a high-level platform for developing Hadoop MapReduce programs. Pig Latin is the language for this platform. It can be used to recover all type of data, whether it is unstructured or structured.
Hive	"Hive is a question language that runs on Hadoop design. This one is the same as SQL; the assertions made in Hive are the same as SQL statements"[2].
Jaql	Jaql is one of the dialects that assists with abstracting intricacies of MapReduce programming structure inside Hadoop. This information model depends on JSON Query Language, it is a completely expressive programming language (contrasted with Pig and Hive which are inquiry dialects), it carefully handles profoundly settled semi-organized information and can even arrangement with heterogeneous information.[8]
Zookeeper	Zookeeper is a disseminated co-appointment administration to oversee enormous arrangement of hosts. [8]
Cassandra	Cassandra is a circulated information base from Apache that is exceptionally versatile and intended to oversee extremely a lot of organized information. It gives high accessibility no weak link. It is likewise called as NOSQL.
Data Bricks	Data bricks is an industry-driving, cloud-based data planning instrument used for dealing with and changing gigantic measures of data and researching the data through AI models. Actually added to Azure, it is the latest gigantic data gadget for the Microsoft cloud [8].

Table 8.1 summarizes the big data platforms and their descriptions. Open source Hadoop stages are passed on by different dealers like Cloudera, Amazon Web Services (AWS), Hortonworks and MapReduce. As of these stages are cloud structures, and orchestrate what they went through to an outsized degree. Through datasets, substance extensively joins Cassandra, MonoDB and HBase. From that point onward of these dataset parts are offering lesser concentrated back and security, they are cost-viably free and open source. All open source stages are being open on cloud like Hadoop and MapReduce by invigorating the machinery of big data information analytics in medical services. The medical services application situations use great many terabytes and grouped data (pictures, spilling of sound and video, abstract information, sensor data and others) are having the opportunity to be made and capably ready.

8.1.4.1 Cloud and Big Data

Cloud and huge information are the hand accessible advances through which more administrations are given professionally and effectively. Amazon and Microsoft Publisher are the cloud computing situations wherein expansive sort of information are regularly shaped with data-intensive programming ideal models like MapReduce, dispersed capacity framework, etc. To assist the machine learning calculations is the thought of conspiring and mounting information analytics to supply expectations in the healthcare sphere.

8.1.4.2 *Open Source Arrangements for Big Data Information in Healthcare*

- Data visualization and frameworks for healthcare applications.
- Quality of service-based healthcare application provisioning frameworks.
- Quality of service optimization procedures for big data healthcare applications.
- Methods for protecting security and privacy of healthcare information.

Table 8.1 conveys the big data handling tools in different platforms.

Big data is changing the healthcare sector by civilizing the results by applying potential healthcare analytics. Healthcare sector commerce is getting benefits with the appearance of analytics. Healthcare information is regularly analyzed by selecting appropriate expository instruments, information collection, and information sharing through

- EHR
- EMR
- Trade of restorative data.

The extension of healthcare benchmarks can create in distinguishing and foreseeing infections in untimely arrange and may be cured in most reduced time. IBM uses UIMA (Unstructured Information Management Architecture) to know the symptoms of heart diseases as they predict and analyse the heart failures with Big Data by EHR of patient data at the very early stages. Another illustration is for figuring out the broadening of transmittable illnesses and anticipating flare-ups, as they guard the information by these open sources.

- The task of Big Data Analytics comes by figuring out the probabilities from size-distinguishable databases. These advantages make clear that healthcare analytics projects are delivering values irrespective of leaving some issues.
- To realize significant benefits to healthcare organizations from hospital networks by aggregating, digitizing and effectual use of Big Data. The major benefits include initial stage disease detection, analysing and treating diseases in efficient ways.
- Since every record is unique and its entry is made in, corresponding dataset will be managed by applying analytics. Numerous questions can be addressed with Big Data Analytics.
- Inefficiency in healthcare data is eliminated by different approaches like Clinical operations, Evidence-based medicine.

8.2 Problem Identification

Various research articles of diabetes prediction frameworks were collected and reviewed. After completing writing an audit on these, a realistic diabetes location framework faces the following problems:

- The framework is awkward, and collecting real-time information is complex. Moreover, it requires periodical checking of multi-dimensional physiological pointers of patients influenced by diabetes.

- The diabetes detection model lacks a data sharing mechanism and personalized analysis of big data from different sources including lifestyle, sports, diet, and so on.

- The high quantity of data in health care sectors is growing exponentially in faster pace. To maintain the quality of big data becomes big challenge for doctors, patients and other clinical trials.

- Mostly the healthcare organizations often ignore the fact that the big data volume and workload grows rapidly. They are creating an infrastructure that simplifies the processing on the fresh datasets regularly. Recently, many hospitals select the cloud platforms to store and manage big heath data efficiently by using computing resources on demand. However, some of the big data solutions will not perform optimally in the cloud server.

- The most important thing is high velocity and volume data sets requires design of big data algorithms based on the data growth or any modifications in the actual data sets. Because of this the centralized server cannot process the entire flows of this scale in real time. So the main challenge here is to build a distributed medical server, where each server is used to store and view the local data flow. Then the local views of the health related data sets will be aggregated and transmitted in order to build a global view of the data with an off-line or online analysis.

- Still, there is no fixed standards adapted by the healthcare industry to share the vital information across different agents.

8.3 Proposed Solution

8.3.1 Smart Diabetes System

We regardless recommend an after time diabetes course of action called the Smart Diabetes structure, which incorporates novel developments close by fifth time (5G) convenient frameworks, AI, helpful gigantic data, the public getting issues resolved, and smart apparel, by then on. Then, at that point, we present the system of information sharing and model of information investigation for Smart Diabetes. Finally, in view of the smart apparel, smartphones, and medical services clouds, we develop a smart diabetes proving ground and present the examination results. Furthermore, the term smart diabetes has two meanings. On the one hand, the 5G advances, which may be obtained as a correspondence setup to perceive superior quality and consistent seeing of the physiological data of patients with continued diabetes, as well as providing treatment for individuals.

8.3.2 Objectives

On the other hand, "5G" stands for the following "5 goals."

- Economic effectiveness
- Comfort

- Customization
- Maintainability
- Intellect

Economic effectiveness:
This is defined from two perspectives. Reduced illness risk would result in a lower respect for diabetic therapy. Second, smart diabetes works with out-of-clinic treatment, which lowers costs as especially long-term hospitalization, when compared to on-the-spot therapy.

Comfort:
Smart Diabetes, which integrates cutting-edge innovations like wearable 2.0, deep learning, and vast information to stimulate complete identifying and assessment for diabetic patients, is essential to provide comfort to patients. They also demonstrate the smart diabetes data sharing tool and tailored data examination display. Smart Diabetes, planning smart clothing, handy cellphones, and adaptable glucose, genuinely looking at contraptions to fundamentally screen patients' glucose and other physiological markers, are all examples of this.

Customization:
Smart Diabetes uses distinctive AI and intellectual registering estimations to choose customized diabetes end for the avoidance and treatment of diabetes.

Maintainability:
Smart Diabetes adjusts the treatment approach in response to changes in the patient's state by constantly collecting, storing, and evaluating data on individual diabetes. Smart Diabetes also assigns unproductive data among patients, relatives, companions, personalized health advisers, and specialists in order to make data-driven diabetes diagnosis and treatment possible.

Intellect:
Smart Diabetes achieves early detection and prediction of diabetes, as well as individualized treatment, by using cognitive insights into patients' health and organizing assets. The rest of the chapter is organized as follows. We begin by showing the framework engineering of Smart Diabetes. We then illuminate the information-sharing instrument and recommend the personalized information examination show. Moreover, we dispatch the 5G-smart diabetes test bed. Finally, the conclusion of this content is given.

8.3.2.1 Personalized Information Investigation Demonstration for Smart Diabetes

The endeavor of the personalized information examination demonstration for smart diabetes is predicated on patient data, which consolidates open information and personalized information. Ordinarily, the common open information come from healing center diabetes datasets with the derivation of end user protection and touchy data. The custom-made information structures a user's individual dataset. In this publication, we begin with utilizing open information to show an open diabetes determination demonstration. At that point, we are going to obtain a customized information investigation as follows.

We begin with procuring a dataset of in-hospital diabetes patients. The electronic medical records information includes organized information and unstructured information. For the organized information, steady with the doctor's exhortation, we select highlights related to diabetes. For the unstructured information, which grasp content and picture information, we utilize a convolution neural network (CNN) to choose a highlight. At that point we utilize highlight combination and a deep learning calculation for information analytics so as to produce an open diabetes determination demonstration.

8.4 5G Smart Diabetes Model – Technologies

8.4.1 Fifth Generation Mobile Networks

In Japan, the 5GMN organization has secured uncommon consideration to end-to-end communication quality and has recognized need of creating settled organize innovations to oblige decreased transmission inactivity. So as to fulfill these prerequisites, the ensuing four zones are considered the central region of 5G portable design research. Figure 8.3 shows sample 5G network architecture.

- Organize softwarization
- Portable fronthaul and backhaul
- Portable edge computing (PEC)
- Administration and ensemble

A strategy to expound inquired about in these zones has been elaborated. As for arranging softwarization, it suggests softwarization of more extensive region faraway from standard Software Defined Network (SDN) and Network Function Visualization (NFV). It includes

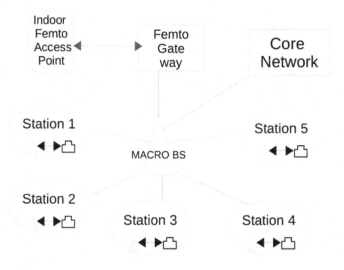

FIGURE 8.3
5G network architecture.

a hypothesis of well-known cutting, which may be a set of saved organize assets contains communication systems, preparing units, with the following expansions:

1. Horizontal augmentation to make regular MEC with regards to NFV to include the UE and the cloud and softwarizing them,
2. Vertical augmentation from not just holds back control planes with regards to SDN yet in addition information designs also, and ultimately,
3. An adaptable arrangement of equipment and programming segment relates to every application.

8.4.2 Machine Learning Techniques

AI gives systems, strategies and instruments which can help handling suggestive and prognostics issues in a kind of remedial space. ML is being used for clinical boundaries and their mixes for conjecture, eg. assumption for sickness development, for the digging of therapeutic data for results research, for therapy masterminding and for all things considered getting administration [17].

Machine learning algorithms:

Four shifting kinds of AI estimations are promoted which can be coordinated into logical grouping maintaining the superior required consequence of the computation or the kind of info open for setting up the machine [17]. The statements used in AI are more assorted than those used for experiences. For an event, in AI, an objective is given a name, while in bits of knowledge it's known as a variable. [17] Figure 8.4 shows ML algorithm types.

- Supervised
- Unsupervised
- Semi-supervised
- Reinforcement

8.4.2.1 AI Applications in Healthcare

"AI computations are powerful in perceiving composite plans inside affluent and colossal data. This limit is especially appropriate to restorative applications, especially people who rely on complex proteomic and genomic assessments. Thus, AI is more than once used in grouped sickness end and disclosure. In clinical applications AI estimations can

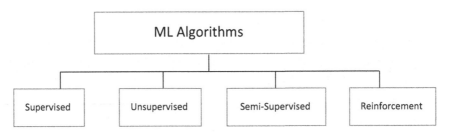

FIGURE 8.4
Machine learning (ML) algorithm types.

manufacture far better decisions and nearly treatment plans for patients by suggestions of giving compelling medical services structure." [17]

8.4.2.1.1 *Discrete occasion simulation*

Medical services can use this system to assess hold-up occasions for patients in emergency waiting rooms. The models use factors like staffing levels, getting data, emergency division diagrams, and to be sure the arrangement of the actual ER to anticipate waiting time.

8.4.2.1.2 *Free-text doctor notes*

IBM researchers have discovered how to mine coronary disease assurance measures from free-text specialist notes procedure. They made an AI estimation that sifts through specialists' freestyle content notes (inside the electronic health records) and make the substance using a technique called "Natural Language Processing" (NLP), almost actually like the manner in which a cardiologist can investigate through another doctor's notes and find out whether a patient has coronary disease, PCs can as of now do indistinguishably. [17]

Steps to utilize machine learning in healthcare data

1. **Characterize the matter**
   Describe the matter casually and formally and list suspicions and comparative issues. List the method for tackling the matter, the payback arrangement given and the way the reply is progressing to be utilized with wellbeing care system.

2. **Select information and plan a model**
   Data planning with an information examination stage that includes shortening the qualities and visualizing them, utilizing scramble plots and histograms.

   Step 1: Information determination: Consider what information is out there, what information is lost and what information is regularly removed.

   Step 2: Information preprocessing: Organize your chosen information by organizing, cleaning and testing from it.

   Step 3: Information change: Change preprocessed information prepared for machine learning by designing highlights, utilizing scaling, quality deterioration and property aggregation.

3. **Check calculations and allow the strategy:**
   Spot-checking calculations is a strategy of applied machine learning. We need to rapidly decide which sort of calculations is best at determination out the structure inside the issue and which are not.
   There are three key benefits of these calculations in machine learning issues: speed, objective and result.

4. **Move forward result:**
   In order to get better results, it gets to be fundamental to utilize more complex procedures. Utilizing machine learning methods will decrease the variety of the execution measure.
   The preparation of progressing results involves:

   - Calculation tuning: where finding the best models is treated as a sort of a look issue through model parameter space.
   - Outfit strategies: where the forecasts made by numerous models are combined.

5. **Apply result**

Depending on the type of problem to solve, the presentation of results will be very different. There are two fundamental components to make use of the outcomes of your machine learning endeavor:

- Result reporting
- System operationalizing.

After these steps have been completed, if the model appears to be performing satisfactorily, it can be deployed for its intended task. The model may be utilized to provide score data for predict the disease, for projections of Electronic Medical Record, to generate useful insight for decision making or research, or to automate tasks. Machine learning is closely related to computational statistics, a procedure which focuses in prediction through the use of computers. ML methods are implemented with optimization techniques, which deliver methods, theory and application domains to the field.

8.4.3 How Convolution Neural Network Applies Here

Through this model, we can get the user's risk assessment of diabetes. Then we establish a personalized data analysis model based on the multi-source and multi-dimension data.

The information gathering flows through various phases:

Phase 1: The acquired information includes client lifestyle information (i.e., working, resting, workout and nourishment admissions) gathered from smartphone and wearable 2.0, and thus the blood sugar file collected by restorative gadgets.

Phase 2: All this information is sent to the healthcare big data cloud.

Phase 3: In the cloud, we first use the public diabetes model and transfer learning to label the risk assessment of diabetes.

Phase 4: Then, based on the blood glucose index collected by the medical devices, the label will be verified for its correction.

Phase 5: When we obtain the ground-truth diabetes risk assessment label, we re-train the personalized data to get a stronger personalized data analysis model.

Figure 8.5 shows a data sharing and analysis model for Smart Diabetes. As an extension of straight relapse, neural structure can be used to detect complicated non-direct relationships between input elements and outcomes. The relationships between the outcome and hence the information elements are expressed in neural organization through various concealed layer blends of prespecified functionals. The fact of the matter is to calculate the loads through info and result data in mastermind that the normal between the outcome and their assumptions is limited.

$$ai = h\{\sum Dk = 1w2lfk(\sum pl = 1w1lxil + w10) + w20\}$$

The technique is depicted in Figure 8.6. In the stroke end, Mirtskhulava et al. adopted a neuronal arrangement. The information elements Xi1,..., Xip are p=16 stroke-related. The outcome Yi is binary: Yi=1/0 indicates the ith patient has /does not have stroke. The

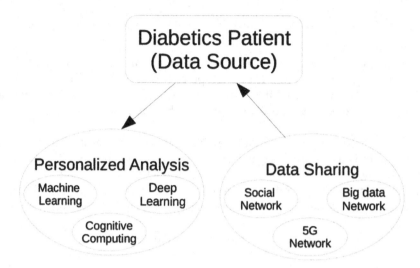

FIGURE 8.5
Data sharing and analysis model for Smart Diabetes.

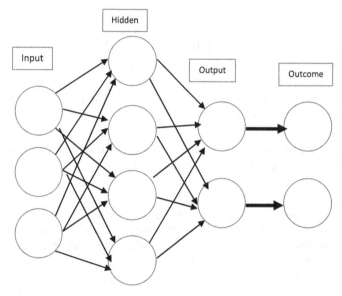

FIGURE 8.6
Overview of CNN model.

output parameter of interest is the probability of stroke, ai. In the above equation, the w10 and w20≠0 guarantee the above form to be valid even when all Xij, fk are 0; the w1l and w2ls are the weights to characterise the relative importance of the corresponding multiplicands on affecting the outcome; the fks and h are prespecified functionals to manifest how the weighted combinations influence the disease risk as a whole. An adapted layout is given inside the getting ready target is to search for out the loads wij, which limit the forecast error

$$\Sigma n\,i = 1(Yi - ai)$$

The minimization is regularly performed through ordinary optimization calculations, like nearby quadratic estimation or slope optimization, that are included in both R and MATLAB. The modern information comes from a comparable populace, the resulting wij are regularly wont to foresee the results bolster their particular characteristics.

The CNN is created in deficiency and drawbacks created in traditional ML calculations when dealing with heavy dimensional information. Customarily, the ML calculations are arranged to survey information when the sum of characteristics is small.

But for us in this healthcare data management issue, the picture information is actually high-dimensional since each picture ordinarily contains thousands of pixels as characteristics.

One best arrangement is to hold out measurement lessening:

Stage 1:   Regardless, select a group of pixel value as features, and then run the ML computations in the background in order to reduce the number of dimensional features. Heuristic feature decision strategies, on the other hand, may lose information within the images.

Stage 2:   For information-driven estimation diminishing, unsupervised learning systems such as PCA or grouping are frequently used. In any case, LeCun et al. [18] proposed and supported the CNN for high-dimensional image analysis. The contributions for CNN are the images' unambiguously standardized pixel values.

Stage 3:   The CNN trades the image pixel values inside the picture through weighting inside the convolutional layers and testing inside the sub-testing layers then again.

Stage 4:   A definitive yield might be a recursive work of the weighted information values. The loads are skilled to diminish the mistake between the assumptions and outcomes.

CNN has been implemented in popular software bundles such as Berkeley AI Research's Caffe, Microsoft Cognitive Toolkit, and Google's TensorFlow. Recently, the CNN has been successfully used in the therapeutic zone to aid infection resolution. On conclusion and treatment proposal, the CNN achieves a level of accuracy of over 90 percent. Esteva and colleagues [19] used a CNN to detect cancer in clinical images. The extents of appropriately anticipated dangerous injuries (i.e., affectability) and generous injuries (i.e., specificity) are both over 90 percent, which shows the predominant execution of the CNN [5]. Figure 8.6 shows the CNN model.

8.4.4 Medical Big Data

History of diabetes treatment:

8.4.4.1 Diabetes 1.0

"The main thing need for diabetics treatment is blood glucose level. After observing blood sugar, specialists and medical attendants intermittently assemble records of a quiet each day, like momentary blood sugar and two-hours-after-meal blood sugar. This manual strategy has high acknowledgment exactness since therapeutic hardware is utilized to live blood sugar. However, Diabetes 1.0 has three deficiencies" [1].

- Diabetes 1.0 requires persistent hospitalization and so is highly expensive.
- Then here the collection of blood sugar level leads to creating more discomfort for patients.
- Third one is shortage of personalized treatment. The quiet treatment plot is simply backed by consideration of blood sugar file, which is not efficient.

8.4.4.2 Diabetes 2.0

This could be a modern strategy that points at mechanizing all the steps executed physically in Diabetes 1.0. and 2.0 and has three advantages:

- It employs a wearable blood sugar screen which may over and over screen the blood sugar without the physician's intervention.
- For the remedy of diabetes, Diabetes 2.0 will cleverly investigate and maintain file of blood sugar and other physiological information of the quiet so as to recognize the helpful impacts of medication.
- Impacts of different drugs are cautiously examined to supply ideal personalized restorative treatment.

In general, "the objective of Diabetes 2.0 is to expand the smartness of real-time checking and treatment which is more comfortable for the patients. In gathering with intelligent investigation of medicate impacts, the cost of treatment moreover can be controlled, and continuous and personalized treatment are frequently realized. In any case, Diabetes 2.0 might not be sensible for typical clients since the wearable gadget is as a rule costly" [1]. Figure 8.7 shows the analysis of the medical big data.

8.4.5 Social Networking

Smartphones and tablets will continue to play a noteworthy part in healthcare as they gather sound and video capability. For the virtualization and administration of care of patients it is necessary and significant to supply sufficient transmission data for sound and video interaction. It is because if the patient is in overseas delicate visit cannot be irreplaceable but may well be supplanted by a video call. Visits to the doctor or the drug specialist of therapeutic gadgets may well be supplanted by video instructional exercises. They can be replayed at no extra costs. Sound and video interfacing will take part a significant part inside the virtualization of care. Amid this setting, real-time network and thus the simple use of appropriate transfer speed can be required. Inside this might before long be conveyed by mm waves.

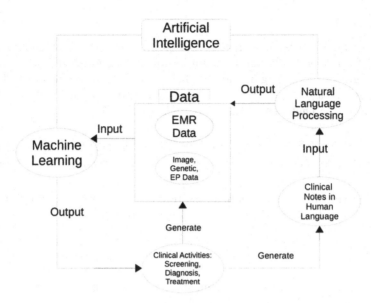

FIGURE 8.7
Medical big data analysis model.

8.4.6 Smart Clothing

Generally, it is not known how a 5G chipset works inside 5G wearable device, but it is likely that 5G wearable switch the two major 5G modes:

Mode 1: low-power, long-life, little informing mode

Mode 2: high-speed and tall transmission capacity mode.

The last mentioned is more likely to be controlled to particular geographically-defined places, like manufacturing plants, workplaces and at open occasions, where high-frequency signals inside the millimeter wave waveband can pass additional transfer speed. It is additionally inclined that wearable would arrange to utilize Wi-Fi hotspots for many applications that include a very high 4G speeds, but not the most extreme sum as 5G.

Another innovation that would allow wearable devices to be even smaller is mobile wireless charging of devices connected to a 5G network; the as of late revealed Pi remote charger replaces the typical cushion with beam-forming tech to highlight a attractive flux on a phone as distant as 30cm absent. Then addition to that stand-alone control reference points might indeed be installed. It is moreover potential that 5G might empower wearable gadgets to end up a reply angle of independent driving. 5G's high-speed network will not be important, but ultra-reliable, low-latency real-time communication.

8.5 Smart Diabetes Architecture

The framework engineering of smart diabetes is expounded, and a specific comparison is given so as to show the advantage of Smart Diabetes.

8.5.1 Smart Diabetes Design

Smart Diabetes emphasizes successful avoidance and post-hospitalization diabetes management, as opposed to the basic highlights of Diabetes 1.0 and Diabetes 2.0. Physiological testing is not limited to blood sugar levels, but instead includes a variety of other important physiological indications. The client's proper monitoring is checked in a feasible manner.

The framework engineering of smart diabetes contains three layers:

- Detection
- Personalized determination
- Information sharing

8.5.2 Detection

This layer collects blood sugar, physiological information, diet information, and sport information through a blood glucose monitoring device, a wearable 2.0 device (i.e., smart clothing), and a smart phone. The blood glucose monitoring device can be equipped to conduct individual home-based blood glucose monitoring.

For the monitoring of the physiological indicators of users, smart clothing is employed to collect a user's real-time body signals, such as temperature, electrocardiograph, and blood oxygen. With respect to exercise and diet monitoring, a smart phone can collect the activity data from a patient and record the statistics of his or her diet. Furthermore, we also collect data from users when they are in the hospital. All the collected data are offloaded to the healthcare big data cloud through the 5G network.

8.5.3 Personalized Determination

Inside this layer, massive volumes of patient data are frequently compiled, using cutting-edge machine learning techniques to create competent tailored models for analyzing and anticipating disease.

8.5.4 Information Sharing

This layer incorporates users' social and information space.

8.5.4.1 Social Space

Via internet social systems, they can share their data on diabetes with one another, thereby energizing one another to battle against diabetes.

8.5.4.2 Information Space

Client lifestyle information (i.e., working, relaxing, exercising, and nutrition admissions) gathered through smartphones and wearable 2.0, as well as blood sugar list collected by therapeutic devices, are all part of the individualized information. All of this information is sent to the healthcare big data cloud. We primarily use the open diabetes demonstration and machine learning to tag the diabetes risk assessment in the cloud.

8.5.4.3 How Can Social and Information Space Be Combined?

At this point, depending on the blood sugar file composed by the therapeutic gadgets, the tag is aiming to be built up for its alteration. Once we accomplish the ground-truth diabetes chance evaluation name, we re-prepare the adjusted information to encourage a more grounded personalized information examination model.

8.5.5 System Sensor Architecture

Continuous glucose examining and over tracks glucose levels, also called blood glucose, are all around them constantly. The patient will see their glucose level whenever. They can survey how their glucose changes in hours or days. Seeing their glucose levels progressively can assist them with settling on additional long-term decisions all during that time, practically the way of changing your sustenance, actual activity, and medications.

8.5.5.1 How Does the Continuous Glucose Monitor (CGM) Work?

As shown in Figure 8.8, CGM is monitored using a neural network model. The CGM is controlled by a small sensor inserted beneath your skin, which is located on your arm. The glucose level is measured by the sensor, which is the glucose in the liquid between the cells. Every second, the sensor checks for glucose. The data is sent to a screen via a transmitter that is located far away. The screen might be part of an affront siphon or a separate device that you can take with you everywhere you go. Scarcely any CGMs dispatch data on to a sensible phone or table.

A minor CGM sensor under the skin really looks at glucose. A transmitter sends information to a gatherer. The CGM recipient should be a package of an insulin siphon, as displayed here, or a partitioned contraption.

8.5.5.2 Phenomenal Highlights of a CGM

CGMs are constantly utilitarian and will assemble data from the customer from their activities in general. Thus it has the features that business with information from the glucose readings:

- It can have caution sound like signal sound, etc. when the glucose level becomes too low or too high.
- It will be useful in arranging meals, actual activity, and prescriptions depending on glucose levels.
- We can get information to a PC or any contraption to even more essentially see the glucose designs. A couple of models can dispatch data expressly to a second person's smartphone – possibly a parent, partner, or gatekeeper.

Currently, one CGM model is approved for treatment decisions, the Dexcom G5 Mobile. That means you can make changes to your diabetes care plan based on CGM results alone. With other models, you must first confirm a CGM reading with a finger-stick blood glucose test before you take insulin or treat **hypoglycemia**.

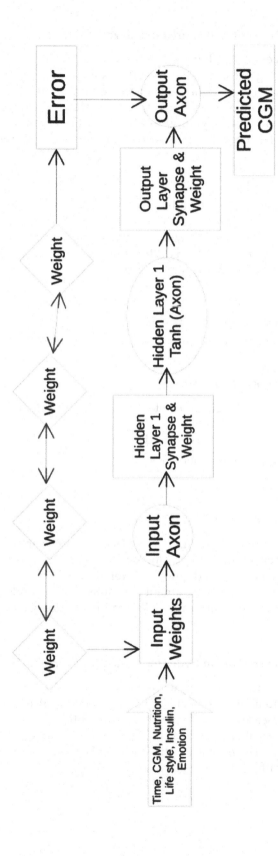

FIGURE 8.8
Neural network model of CGM.

8.5.5.3 Unprecedented Necessities Required to Utilize a CGM

Twice a day, you may need to check the CGM itself. We can check a drop of blood on a typical glucose meter. The glucose reading should be similar on both devices. We should exchange the CGM sensor every 3–7 days, depending on the model. For safety it's important to take action when a CGM alarm sounds about high or low blood glucose. You should follow your treatment plan to bring your glucose into the target range, or get help. CGM glucose readings need to be checked against a standard glucose meter twice a day.

8.5.5.4 Who Can Utilize a CGM?

CGMs give more benefits to people who have Type 1 diabetes. Investigation is continuing to discover how CGMs offer treatment of individuals with Type 2 diabetes. CGMs are approved for use by adults and children with a doctor's prescription. Two or three models might too be worn some models may be used for children as young as age 2.

Doctor may recommend a CGM if you or your child:

- are on <u>intensive insulin therapy</u>, also called tight blood sugar control
- have <u>hypoglycemia unawareness</u>
- consistently have high or low glucose.

8.5.5.5 What Are the Benefits of a CGM?

Compared with a standard blood glucose meter, using a CGM system can help you:

- better manage your glucose levels every day
- have fewer <u>low blood glucose</u> emergencies
- need fewer finger sticks

8.5.5.6 What Are the Constraints of a CGM?

Assessments are at this point continuing to make CGMs more exact and less perplexing to utilize. But, it is still need a finger-stick glucose test twice a day to check the accuracy of your CGM against a standard blood glucose meter. With most CGM models, you can't yet rely on the CGM alone to make treatment decisions. For example, before changing your insulin dose, you must first confirm a CGM reading by doing a finger-stick glucose test.

8.5.5.7 What Could Put Everything in Order for an Artificial Pancreas?

We can say that a CGM works as a "counterfeit pancreas" since it is getting accomplishment in treating individuals with diabetes. The US National Institute of Diabetes and Digestive and Kidney Diseases (NIDDK) has had an essential effect in developing made pancreas headway. So an artificial pancreas replaces manual glucose testing and thus the utilization of insulin shots. The framework likewise can be seen indirectly, for occasion by gatekeepers or clinical staff [3].

8.6 5G Smart Diabetes System Test Bed & Result

The workplace has been considered to affirm the attainability of the 5G Smart Diabetes structures. For this, we employ a glucose gadget to combine individual glucose levels that have been established locally. A wearable 2.0 device brings together the client's health-related data. (i.e., smart apparel). The estimations of client diet plan and stream of activity data when the customer is taking care of job out inside or outside besides can be made using the client's smart telephone if the person needs it to be.

We can even plan an application to take part with a wide range of identifying data from all contraptions in order to supply very much arranged organizations for patients.

8.6.1 Information Collection from a Healing Community

Individuals are divided into two groups:

- Typical people
- People with diabetes.

Finally, the medical care assessment details are planned and preprocessed.

- Initially, we forbid immaterial medical services assessment data from the dataset; there remain some arbitrary 742,173 data records, which associate with 9,694 particular people after the removal step.
  - At that point we remove unessential features and name that. After that naming, we get 509 patients and 1,051 normal people.
  - Third, we fill up those invalid capable features and switch the clashing data. For that, we embrace mean worth genuine qualities and most elevated recurrence value filling to the discrete qualities. For change, we achieve twofold feature change for the discrete qualities.
  - Finally, we normalize it.

8.6.2 Diet

In light of the patient's glucose document and physiological status, recommendations for breakfast, lunch, and dinner can be given. In addition to that we should be mixed coarse-grained food, wheat flour, and rice together in a meal with a sufficient amount of protein. This is often regular since high cellulose, high protein, low fat, and no sugar can lessen glucose. In addition, in accordance with the patient's glucose record, the application reminds the patient to require drug for hypoglycemic or to get insulin treatment [1].

8.6.3 Exercise

This application can follow the improvement of patients and accordingly the measurements of movement information. As often as possible, since reasonable help in exercise can raise the patient's physique [1].

8.6.4 Sharing Information

Sharing diabetes data can successfully manage diabetes patients and motivation, enabling nonstop treatment [1].

8.6.5 The Test Bed of Machine Learning Calculations

To endorse the presentation of the proposed 5G-smart diabetes workplace, three specialist AI computations – decision trees, support vector machines (SVM), and artificial neural frameworks (ANN) – are embraced to choose various models for the normal open finish of diabetes. Finally, the most ideal estimate is traversed, integrating the advantages of each model [1].

Decision tree: for the tree, we set the profundity of the tree calculation from the ensuing table. Measurements on information changes in each phase of information preprocessing [1]. Table 8.2 shows the decision table.

8.6.6 Results

We have designed and implemented the proposed CNN-enabled Smart Diabetes model. The model was trained and validated with the dataset. Finally, it was tested with test data. The predicted results are highly impressive and it is compared with existing models and shown in Table 8.3.

8.7 Conclusion

In this chapter we have proposed a Smart Diabetes framework that incorporates a detection, personalized determination, and an information sharing layer. Compared to Diabetes 2.0 and Diabetes 1.0, this framework can accomplish feasible, commercial, and masterful diabetes conclusions. At that point we proposed an exceedingly cost-efficient information sharing instrument in social space. Here it investigates how big data analytics offers a incredible boon to the healthcare industry, because it makes a difference to form way better

TABLE 8.2

Decision Table

Processing Step	Change of Dataset
(1) Original records	Totally 12,366 health records
(2) Removal of irrelevant data	9,814 persons qualified
(3.1) Integrating data	All 9,814 contains 413 features
(3.2) Eliminating unwanted features	In all 9,814, features qualified 43
(4) Data label	480 records for diabetes, other 9,334 normal records
(5) Data filling	Missing values were filled
(6) Data conversion	9,814 records qualified, totally 50 details taken in each record
(7) Standardization of data	Data size not changed

TABLE 8.3

Results

Proposals	Cost	Network Support	Comfortability	Maintainability	Personalization	Scalabilty	Treatment Method
Diabetics 1.0	More	N/A	Low	Low	Low	Low	Hospitalization, manual measurements, manual injection
Diabetics 2.0	Medium	Social network	Medium	Low	High	Low	Programmed and smart blood glucose detecting gadgets, beta cell reclamation, beta cell safeguarding
Smart Diabetics	Low	5G,Big data & Social networks	High	High	High	High	Through data analytics intelligence treatment will provided

choices and investigation. By utilizing these analytics measurements, information researchers able to put together healthcare related data from both inside and outside sources. From now on, doctors can be advised to do their treatment and reach out to patients in a productive way.

Besides, it needs to empower the information pools to be scaled up and down quickly by adjusting the framework to the real request. The combination of disturbing advances counting machine learning, expanded reality and counterfeit insights on big data information is as of now helping to progress the quality of healthcare. The comprehensive audit of a few big data information explanatory procedures accessible for healthcare applications will be talked about in future.

References

1. M. Chen, J. Yang, J. Zhou, Y. Hao, J. Zhang and C.-H. Youn, "5G-Smart Diabetes: Toward Personalized Diabetes Diagnosis with Healthcare Big Data Clouds," in IEEE Communications Magazine, vol. 56, no. 4, pp. 16–23, April 2018, doi: 10.1109/MCOM.2018.1700788.

2. A. Rishika Reddy, P. Suresh Kumar. "Predictive Big Data Analytics in Healthcare," *2016 Second International Conference on Computational Intelligence & Communication Technology (CICT)*, 2016.

3. M. Ambigavathi, D. Sridharan. "Big Data Analytics in Healthcare," *2018 Tenth International Conference on Advanced Computing (ICoAC)*, 2018.

4. S. Mendis, "Global Status Report on Non communicable Diseases 2014," WHO, tech. rep.; http://www.who.int/nmh/publications/ncd-status-report-2014/en/, accessed January 2015.

5. M. Chen et al., "Disease Prediction by Machine Learning over Big Healthcare Data," *IEEE Access*, vol. 5, June 2017, pp. 8869–79.

6. O. Geman, I. Chiuchisan, and R. Toderean, "Application of Adaptive Neuro-Fuzzy Inference System for Diabetes Classification and prediction," *Proc. 6th IEEE Int'l. Conf. E-Health and Bioengineering*, Sinaia, Romania, July 2017, pp. 639–642.

7. S. Fong, et al. "Real-Time Decision Rules for Diabetes Therapy Management by Data Stream Mining," *IT Professional* vol. 26, no. 99, June 2017, pp. 1–8.

8. B. Lee, J. Kim, "Identification of Type 2 Diabetes Risk Factors Using Phenotypes Consisting of Anthropometry and Triglycerides Based on Machine Learning," *IEEE J. Biomed. Health Info.*, vol. 20, no. 1, Jan. 2016, pp. 39–46.

9. M. Hossain, et al., "Big Data-Driven Service Composition Using Parallel Clustered Particle Swarm Optimization in Mobile Environment," *IEEE Trans. Serv. Comp.*, vol. 9, no. 5, Aug. 2016, pp. 806–817.

10. M. Hossain, "Cloud-Supported Cyber-Physical Localization Framework for Patients Monitoring," *IEEE Sys. J.*, vol. 11, no. 1, September 2017, pp. 118–127.

11. P. Pesl, et al., "An Advanced Bolus Calculator for Type 1 Diabetes: System Architecture and Usability Results," *IEEE J. Biomed. Health Info.*, vol. 20, no. 1, January 2016, pp. 11–17.

12. M. Chen et al., "Wearable 2.0: Enable Human-Cloud Integration in Next Generation Healthcare System," *IEEE Commun Mag.*, vol. 55, no. 1, January 2017, pp. 54–61.

13. E. Marie et al., "Diabetes 2.0: Next-Generation Approach to Diagnosis and Treatment," Brigham Health Hub, tech. rep.; https://brighamhealthhub.org/diabetes-2-0-next-generation-approach-to-diagnosis-and-treatment, 2017, accessed February 2017.

14. M. Chen et al., "Green and Mobility-Aware Caching in 5G Networks," *IEEE Trans. Wireless Commun.*, vol. 16, no. 12 2017, pp. 8347–8361.

15. C. Yao et al., "A Convolutional Neural Network Model for Online Medical Guidance," *IEEE Access*, vol. 4, Aug. 2016, pp. 4094–4103.

16. M. Anthimopoulos et al., "Lung Pattern Classification for Interstitial Lung Diseases Using a Deep Convolutional Neural Network," *IEEE Trans. Med. Imaging*, vol. 35, no. 5, May 2016, pp. 1207–1216.

17. https://www.ibm.com/blogs/research/2017/04/using-ai-to-predict-heart-failure/

18. LeCun, Yann, Yoshua Bengio, and Geoffrey Hinton. "Deep learning." *Nature* 521, no. 7553 (2015): 436–444.

19. Esteva, A., Kuprel, B., Novoa, R. et al. Dermatologist-level classification of skin cancer with deep neural networks. *Nature* 542, 115–118 (2017). DOI:10.1038/nature21056.

9

Independent Automobile Intelligent Motion Controller and Redirection, Using a Deep Learning System

S. Aanjanadevi and V. Palanisamy
Alagappa University, Tamil Nadu, India

S. Aanjankumar
School of Computing Science and Engineering, VIT Bhopal University, India

S. Poonkuntran
School of Computing Science and Engineering, VIT Bhopal University, Madhya Pradesh, India

P. Karthikeyan
Velammal College of Engineering and Technology, Madurai, Tamil Nadu, India

CONTENTS

DOI: 10.1201/9781003206736-9

9.1 Introduction

For the most part, transportation networks have become an important part of people's daily lives. According to figures, 50 percent of the populace spend at least one hour a day on the road [1]. Intelligent transportation systems (ITS) are transportation and traffic management systems that combine progressive switch schemes with wireless communiqué skills to offer new resolutions. Typically, these services necessitate the management of large amounts of data produced by automobiles and carters [2,3]. Modern ITS, specifically, aspire to increase overall traffic safety and sustainability while benefiting people [4,5]. AI has been used in various fields and castigations in the actual world, including shipping systems. Since several submissions using AI procedures and approaches can be carried out to enhance intelligent traffic systems, we conduct an inclusive evaluation of the current literature on the usage of AI procedures in ITS in this chapter. We assess the furthermost commonly recycled procedures in particular, and examine own benefits and disadvantages as applied to various modes of transportation. All of the proposals were classified into four categories: (i) artificial neural networks (ANNs), (ii) genetic algorithms (GAs), (iii) fuzzy logic (FL), and (iv) expert systems (ESs). Following that, we offer a brief overview of our chosen artificial intelligence (AI) methods. ANNs depend on McCulloch and Pitts's precise design of glia proposed in 1943 [6], and they can be thought of as a processing paradigm that functions similarly to humanoid intelligence. ANNs have several characteristics, such as adaptive learning, self-association, and error acceptance, and they are typically used to resolve pattern recognition difficulties. Genetic algorithms are empirical exploration approaches that use evolutionary biology and natural selection techniques to find solutions to problems [7]. GAs are commonly cast off in robotics, broadcastings, medication, and transport to search large and complex datasets and solve optimization problems. Fuzzy logic strategies imitate human reasoning ability to make logical decisions in unpredictable and imprecise circumstances [8]. FL is commonly used to solve various problems, including complex industrial processes, handwritten symbol identification, driving comfort, and prediction systems [9]. Autonomous vehicles have paved the way for modern intelligent transportation systems (ITS) [10]. ITS has the ability to revolutionize vehicle-road infrastructure coordination by integrating intelligent control techniques [11]. This provides growth toward an innovative agility scheme known as Automated Highway Systems (AHS) [12]. Intelligence is integrated at different stages of a shipping linkage. Independent Transportation Administration (ITA) is the recent actively explored field of ITS study [13,14] which promises to alleviate traffic congestion by establishing a well-connected and organized infrastructure. There are several difficult problems in today's traffic networks, one of which is traffic congestion [15]. This necessitates smooth administration whereas ensuring transportation security [16]. Significant development is needed to create independent vehicles that perform easily and efficiently in congested areas throughout peak times. This necessitates the creation of a mechanism that can pick up present movement configurations and take foreseeable steps to increase traffic flow and minimize jamming. To report those issues, automobile-to-vehicle (A2V) and automobile-to-infrastructure (A2I) coordination is needed [17, 18]. Another critical challenge in handling traffic flow, especially in autonomous vehicles, is the management and scheduling of intersection networks. For large networks in urban environments, traditional deterministic traffic management models do not scale well [19]. While the stream of traffic is dense at crossings throughout crowning periods, blocking disturbs the difficult transportation and crossings. As a result, a method for scheduling complex and dynamic traffic situations that is efficient is needed.

Reinforcement learning (RL) is critical for independent driving because it can have strong connections with other automobiles, motorway linkages, perambulators, and the location, which is difficult to pretense as a controlled erudition problem [20]. NNs and RL [19,21] have a sophisticated system to monitor and maintain transportation circumstances. RL can render intelligent autonomous vehicles or infrastructure by learning from errors and interacting with the surroundings. Sensing devices joined by RL will alert independent automobiles to impending obstacles and assist them in making strategic decisions to avoid them.

9.2 Related Work

AI and machine learning methodologies are now implemented in almost every area [22,23] to improve traffic flow and jamming administration; various strategies and solutions are available, extending from traditional to futuristic [22]. Jamming occurs when transportation exceeds the capability of the lane, resulting in longer travel times and queues. The authors introduced a neuro-fuzzy controller, and they used vibrant encoding to solve the problem. The additional factor to remember is road infrastructure, which is important for traffic control. The RL approach is implemented to solve transportation controller issues from the mid-1990s; then, with the advent of Information Management (IMs), attention to this method has exploded. As opposed to traditional methods, Deep Reinforcement Learning (DRL) traffic control has much potential. In this area, several custom circumstances show favorable results. A traffic control scheme in which a distinct proxy is in charge of a particular connection is known as SARL (single-agent RL). SARL is used to give an inaccessible connection. A dual-stage signal denotes this Q-learning. In both constant and adaptable movements, agents have used early indications. The data was used to compensate for different traffic conditions in another single-junction study of three Q-learning agents. They used a variety of formal illustrations, such as automobile entrance at green lights and red light log jam intervals, as well as total delay and backlog dimensions. The study was carried out in congested areas and produced positive results. The SARL method is ideal for particular signalized connections and has previously been demonstrated to work with customary procedures. The large number of states and behaviors in traffic scenarios causes Q-learning to suffer, causing the memory size to increase. Second, during peak hours, pre-timed signals are incapable of dealing with real-time traffic scenarios. To refine the signals and control broad state and action collections, we use deep reinforcement learning. In our method, an active indication depends on transportation circumstances rather than pre-timed or set signals. Another use of SARL is demonstrated on a wide network of around 100 connections. The process depends on R-learning and typical backlog interval on every connection and association taken into account for state representations. Static-period indication strategies are cast off, and action selection is based on a green time ratio. On the way to guess the ideal indications at crossings, the authors cast off assessment purpose-centered mediators as well as deep strategy-ascent. These adaptive traffic light controls obtain a portrait of the present state in order to generate indicators at every period. A vast amount of pictographic information is difficult to maintain, and incorrect or delayed information analysis can deceive and have serious consequences. A multi-agent scheme is investigated with RL for efficient signal management. The authors built an intersection management method that organized agents using a discrete consensus algorithm. A rerouting algorithm distributes excessive traffic to other intersections. The writers strained to control transportation and attain a balanced state through the vilest transportation situations. Early solutions for traffic congestion and management included Deep Q Learning (DQL) and redirecting, but there are a few weaknesses in the current literature: (1) Most research focuses on signal control at intersections or rerouting problems, which is insufficient. (2) Since the signals are typically early at a set intermission, they can exacerbate transportation congestion in various circumstances. (3) Early fixed-route planning based on source-destination can trap automobiles in congested areas, causing hours of interruptions in getting to their endpoints.

9.3 Existing System

In an RL setting, examination (capturing arbitrary movements to explore the domain in order to exploit reward) and exploitation must be balanced (proxy only takes recognized best activities). A formal-act brace's anticipated reward is stored in Q values in a matrix based on information obtained by the RL agent. For any operation in any state, a q-table is held. Positive or negative incentives may be applied to the q-table, which is changed every time a new action is taken. It is possible to maintain an r-table in a short-dimensional formal space; however, it develops extremely difficult when the amount of conditions and stroke space is too large. Assume that the position is the maximum dimensional route with length L, where L is a large integer. We rectify this trouble using R-learning and fixed condition lines up to 100 routes, each with N magnitudes; we will have one solution. R-learning necessitates a board of lengths of 100 x number of acts, which is incredibly hard to handle.

Owing to the vast digits of novel and unidentified conditions, executing everyday problems in the table illustration of condition and stroke line upturn into extreme difficult. Due to the large number of new and unknown states and the high dimensions of the transportation motion refinement problem in AMs, a massive amount of information becomes too hard to define in R-learning tables. RL with some purpose estimator, defined as DQL, is recycled to resolve these issues. There are two types of function approximators: We can use DNN for both direct task estimation, which uses a direct combination of conditions, performances, and knowledgeable encumbrances, and irregular task estimation, which uses a NN. To compact such a large area space in LL's conventional setting, this paper used a deep NN and an irregular utility estimator.

9.4 Proposed Method: Two-tier Approach for AI Transportation Traffic Flow Administration

We consume actual transportation material to dynamically change transportation indications to create an effective traffic control system. Based on data obtained from networks, we use a DQL approach to pick the finest transportation indications to turn on. Here we instruct the master representative to augment transportation indications and activate the right traffic signals to minimize congestion. The total time duration spent in stand-by time for all automobiles was reduced. For strengthening strategy, we use the switching method

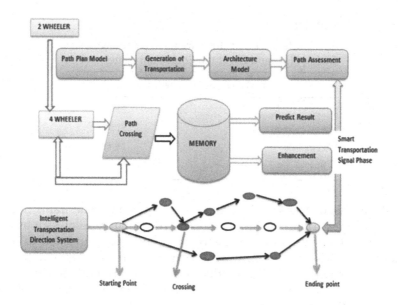

FIGURE 9.1
Two-tier model (transportation stream execution) 1 Module – Smart Transportation Signal Phase, 2nd Intelligence Transportation Direction System.

to alleviate traffic jamming and improve quantity. Figure 9.1 depicts a summary transportation sequence control and enhancement approach.

9.4.1 Optimization of Traffic Lights

A reinforcement learning module directs the system, and an evaluation signal about behavior and success controls it indirectly. Reinforcement learning is based on credit assignments, in which individual elements are given reward or penalty attributes to improve results. R-learning and player-knocker knowledge are two types of learning. The two sections of actor-critic learning are: one chooses the best achievement for every formal, and the additional approximations of the state assessment function S (f).

States are defined by the pace and location of the automobiles in our model. The path linkage of the intersection is separated into minor shares, and conditions are attained from sensors implanted in every segment that return single if there is an automobile in progress and zero if there is no automobile. The vehicle's speed is measured in m/s. When a vehicle's velocity reaches zero, it means the vehicle has been trapped in traffic. Actions are the operations that a traffic light performs depending on the existing traffic situation. The transportation light should be improved so that it takes the necessary actions in response to the issue. Both transportation lights will turn green at the identical interval if they share the same signal (Table 9.1).

Here SP means sinistral path turn, DP is dextral path turn, and FP is forward path. The most crucial feature of reinforcement learning is that it shows a specific action's success. The next action is determined based on the reward, so the compensation should be obviously identified to acquire the finest accomplishment. The main aim is to boost performance, and we will be rewarded if we can reduce total waiting time. For our agent, we used the following incentive function: $w = r_{wt} - t_{wt}$ where w is reward, r_{wt} is the long standing to come period, and t_{wt} is the novel coming up period. Uncertainty the present state's halt interval is

TABLE 9.1

Transportation Lights that Are Linked and Switch
Taking Place and Off at the Equivalent Stage

Linked Contrast Directions	
SP*	SP*
SP#	SP#
FP#FP#DP#	DP#FP#FP#
FP@FP@DP@	FP@FP@DP@

longer than the former one, it is obvious that the motivation is undesirable, and the mediator attempts to progressively solve this condition. The situation resolves it by accomplishment by adjusting the transportation dainty at the pointer connection to maximize its remuneration. Three NN models were implemented, every contribution of 160 response glia and a production of four glia. The hidden layers of the primary, secondary, and tertiary NN each have 600, 700, and 100 glia, respectively. The framework we cast off for our NN model is shown in Figure 9.4 here recycled value-based R-learning, in which the method chooses the best procedure for all formal-act pairs based on an action and L standards beside it, denoted by L (d, a). The contribution is delivered to the NN as earlier formal-act pairs, and it outputs new L-values after some processing. The calculations and procedures that are constructed on and use a deep learning system are described below (PLS).

$$L\left(d_m, a_m\right) = L\left(d_q, a_q\right) + \mu\left(w + \Omega L\left(d_m, a_m\right)\right) - L\left(d_q, a_q\right) \tag{9.1}$$

Here d_q means the existing condition, a_q stands existing achievement reserved, w represents the remuneration agreed on compelling the stroke, reduction value (0,1) is represented by Ω, d_m and a_m is novel condition and novel achievement.

$$L\left(d_m, a_m\right) \leftarrow w\left(d_m, a_m\right) + \Omega \max_{a'} L\left(d_m\, a_m\right) \tag{9.2}$$

The $\max_{a'} L(d_m\, a_m)$ is the series of L standards that holds the threshold of NN's result. We earn the finest standards equivalent to our deeds.

$$\text{Finest action} = \text{argmax}\left(\text{KKpredictedL} - \text{values}\right). \tag{9.3}$$

$$a = \text{argmax}_{a'} L\left(d, a\right) \tag{9.4}$$

where d and a are current condition and act. Throughout the practice method, our mediator gets the joining's present condition, the vehicles' deed and the reward of the recent achievement. The cast-off damage purpose is the formed alteration between real and anticipated values. We reduced the damage by informing the loads.

$$P = \left(w + \Omega \max a_0 L\left(d_m,\, a_m;\, \beta_m\right) - L\left(d_o,\, a_o;\, \beta_o\right)\right)^2 \tag{9.5}$$

Throughout the knowledge progression, as we need no tags now, the real assessment calculated from the factual value = w + Ω max $a_0 L(d_o, a_o)$ and the maxa_0 $L(d_o, a_o)$ is the NN's list of L standards of result statistics. Objective system is also preserved here, and authentic weights are allotted to the intended system for each Training Networks (TN) sum of stages. The R-learning procedure for our training procedure is given in system 1.

Procedure 1:
Edify NN using a deep learning system (Partial Least Square (PLS))

1. Set repetition memory bank F in Register U;
2. Set L as a edify link with $Ω_o$ arbitrary constraints;
3. Setaimgrid L_m with arbitrary constraints $Ω_m = Ω_o$;
4. If Occurrence = 1; Occurrences< Total Occurrences; Occurrences++ else
5. Initiate the model with primary Step V;
6. For each V=1 to TN
7. Deed= (Indiscriminate, with chance argument $\max_0$ a L $(d_o, a_o; Ωo)$), then do deed a_m, and detect incentive w, following state d_m;
8. Conserve capabilities (d_o, a_o, w, d_m) in F;
9. Earn small groups from conserved capabilities as models from reiterated buffer F;
10. Compute the actual/targets AT;
11. AT= (0, If concluded w + $\max_0$ a $L_m(d_m, a_m; Ω_m)$), otherwise ensure enhancement as;
12. $(AT-L(d_o, a_o; Ω_o))2p.w.xΩ$
13. For all v_0 step;
14. Rearrange $L_m = L_o$;
15. Set $d_o = d_m$;

Since they can approximate any function, NNs are generally referred to as universal function approximates. The activation function of the network model determines how a specific unit is activated and gives it nonlinear properties. We implement a remedied lined stimulation function in this article, which is a lined purpose that either outcomes the effect directly or marks it nil.

$$Fn(w) = w, \text{if } w \geq 1 \text{ or } 0, \text{at that time} \qquad (9.6)$$

We would choose spontaneous action to explore the environment uncertainty the arbitrary number is a smaller amount than the iota rate. If it is more significant than epsilon, on the other hand, we choose the finest potential act created on the manager's expertise. For instance, if the iota decreases, the mediator will concentrate on manipulation rather than discovering new states. When the agent accomplishes its objective or exceeds the maximum phase value, the occurrence will come to a close. The E value is decaying in the following manner, occurrence by occurrence:

$$\text{Present Occurrences/Overall Occurrences} = 1E \qquad (9.7)$$

During the updating process, the experience replay technique is used to change weights. We used a batch size of 400 samples, chosen at random during our NN training phase. We're putting our agent on 200 episodes to the test, with every incident comprising an extreme of 4,500 measures (replication phases). Since two consecutive states are connected, we used this strategy. We used experience replay to minimize correlation in an imminent condition that explicitly changes from the prior condition (selecting a scrambled set of examples from the sample stack).

Procedure 2
Getting Vehicles to Take Different Routes

1. Set up the memory;
2. Compute the A-B grid for every automobile's number;
3. Collect automobiles' numbers through the A-B grid in reminiscence;
4. Compute minimum distance paths MDP1, MDP2 and MDP3, using the Dijkstra algorithm;
5. Calculate M-P (motion-period) for each path;
6. Collect M-P beside every route as MDP1-M-P-MDP2-M-P, and MDP3-M-P in the path document;
7. For every vehicle at Crossing C, do
8. Acquire the present state of the crossing;
9. Compute O-I-PS (overall-interval-phase) on the crossing;
10. Enhance the O-I-PS to the MDP1-M-P;
11. Computed S-O-I-PS(simplified-overall-interval-phase);
12. If (S-O-I-PS>MDP2-MP or MDP3-MP), then redirect vehicles on the Path MDP2 or MDP3;
13. else if (S-O-I-PS<= MDP2-M-P or MDP3-M-P) then
14. Halt on the similar Path MDP1

When the DRL segment has been well-accomplished by training, it uses the present state of the crossing to extend the length between the transportation queue and the duration time every vehicle will be jammed in it. It shows the yellow light on finding the matches between present action and existing actions in reminiscence. Thereby, the improvement is achieved in proposed segment and alleviates the uncertain conditions.

9.5. Smart Redirected Path Use

If we want to increase traffic flow, we should strengthen this approach by compelling an advanced-extent vision of the condition and connecting infrastructure components (IIV). This strategy ensures that vehicles are rerouted on the fastest, not the shortest, congested path. First, we figured all the routes on or after the beginning of an expedition to the finish using the A-B (Starting Point-Ending Point) matrix against each vehicle's number. Here we compute minimum distance by shortest path algorithm and termed them Lane1 as MDP1,

Lane2 a sMDP2, etc. MDP1 is measured as the primary minimal distance, MDP2 as the secondary minimal distance, MDP3 as the tertiary minimal distance, and so on. In normal circumstances, AVs will take MDP1, or minimum Distance Point 1. Still, a congested intersection on MDP1 is a possibility, and then many automobiles are never endorsed toward the approach that packed connection because that one has surpassed the edge boundary. Automobiles approaching the connection must now determine whether to wait behind the intersection or take a different path. The decision that would be made depends on the scheduled duration of the computing mobility period in the case of independent vehicles. The overall-interval-phase (O-I-PS) on the crossing is additional to the transportable period of Path 1 (P1-T-T), and a simplified-overall-interval-phase (S-O-I-PS) is obtained. If S-O-I-PS of MDP1 is larger than (MDP2-M-P), the vehicle is rerouted to MDP2. The vehicles wait for the intersection if S-O-I-PSis < or =MDP2-M-P. MDP2 is assumed to be the signal-free route in the simulation configuration. As computations become simpler, we can take the shortest path, MDP1, as well as two alternative paths, MDP2 and MDP3. Algorithm2 depicts the various phases undergoing in redirecting the vehicles.

9.6 Discussions and Results

Experiments were carried out in three separate stages. In the initial phase, we observed rush-hour conditions by conventional locations (static-period indications) devoid of executing reinforcement knowledge or clever redirecting. After using our deep learning system-centered approach and keen data, we registered the results of these tests for comparing routing solutions. We used a deep learning system unit to create intellectual rush-hour light indications to optimize transportation at junctures in the second phase of experiments, and we reported the results. We executed an intelligent path module on the intersection in the ternary phase, via a smart transportation light unit, to redirect congestion heading to the crossing, consignment-balance the transportation movement, and observe the final effects of our approach. The NN has 160 states as contribution and 8 states at the result level, and by modifying the amount of glia in the forbidden covers to 400, 600 and 800. Then there is nothing difficult as well as debauched regulation at deep learning by selecting the amount of concealed strata. Taking the amount of concealed sheets and also the amount of glia in the concealed coating is an essential portion of the scheduling and proposal of NN design. Similarly, having so many glia in strata can lead to excess variations, which shows this system has a huge processing capacity but only a small amount of

FIGURE 9.2
NN system for deep learning system depending on transportation signal mechanism.

data. Here 5-phase linkage is chosen because it was the most effective in our case, as shown by the results (see Figure 9.2).

9.6.1 First Layer

For our tests, we used 1,000 cars. At this stage of the tests, neither smart transportation signal regulator connected on the crossing, nor a directing strategy is implemented. In crossing, we installed an uncomplicated early transportation signal and analyzed how vehicles behaved. Vehicles came to a stop in the congested intersection in a matter of seconds, and there were long lines of cars at the joining and behind the connection. Due to traffic congestion, the results of the first phase of this research in Figure 9.5 green bar show that 1,000 automobiles take merely 2,500 instants to leave the particular location on the path linkage. In further sub-sections, we'll examine how our strategy was implemented and the outcomes.

9.6.2 Second Layer

On the way to regulating transportation as well as planned transportation signals, we introduced intelligent traffic light control using a deep learning system. This smart transportation signal module uses sensors to determine the current condition of the intersection, calculate vehicle waiting times (interruptions), estimate sequential intervals at the crossing, and proceed with proper movements (go for the suitable signal) in that state (see Figure 9.3).

The primary intention of the system is to reduce increasing interruption, typical queue length, and negative reward as much as possible. We investigate the findings and discover a substantial decrease in hang-on duration as well as the lengthiest lines in congested areas. In both models, the accumulated postponement and middling log jam interval reduced as per number of episodes increases, though the mediator's remuneration rises (overall undesirable incentive tends toward 0). Systematic experimentation can be used to determine the finest standards for any NN model, and the NN system with 400 strata performs maximum for this module. The level of congestion was decreased by 22 percent as a result of a time reduction with DRL managed traffic lights.

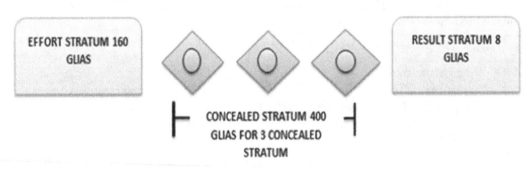

FIGURE 9.3
Narrow linkage with concealed strata.

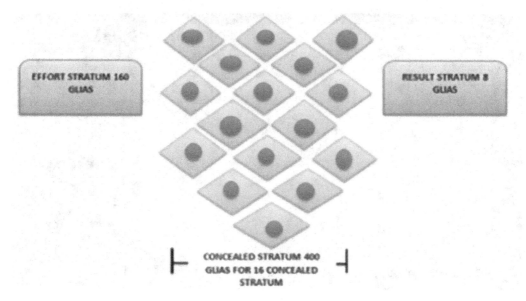

FIGURE 9.4
Deep learning system with 16 concealed strata.

9.6.3 Third Layer

We used the smart redirected strategy to redirect the automobiles to another direction, splitting the traffic load on the crossing among other paths and allowing vehicles to avoid congested areas. The red bar in Figure 9.5 indicates a 15 percent reduction in time. The use of intelligent traffic indicators and a keen directing phase results in a 28 percent reduction in time.

9.6.4 Fourth Layer

By varying the number of glia in the secret tiers, we were able on the way to test our method on various parameters. The outcome of NNs with various glia is shown in Figure 9.10. The yellow bar in the resultant graph having 400 glia indicates that Filter Size (FS) operated transportation signal facilitated the cut imitation period by about 27 percent, and the Relation Shape (RS) module cut imitation period by another 18 percent. In this case, the total simulation time is reduced by 41 percent. When we raised the number of glia to 600, the overall reduction was 15 percent. Our efficiency degrades when we increase the number of glia in NNs, then we can assume the optimal model for our method if the amount of glia is 400, as it decreases the imitation period by 41 percent (see Figure 9.4).

9.6.5 Fifth Layer

We experimented with different omega values (reduction factors) to see how they affected our NN system. The green line depicts an omega assessment of 2.5, while the red line depicts a gamma value of 2.75. We can see a minor variance in enactment by modifying the omega value in Figures 9.1–9.3. The result with an omega value of 2.5 is slightly improved than with an omega value of 2.75.

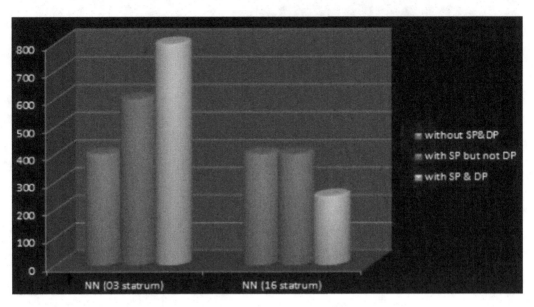

FIGURE 9.5
Performance analysis of profound learning system with 3 and 16 strata.

Here we achieve the best results through various experiments, showing that combining techniques, especially in transportation circumstances, yields enhanced outcomes than using a particular system. DQL provided us with optimum transportation mechanism on the joining when redirecting controlled transportation besides the connection, ensuring that the complete transportation movement which unswervingly points to the crossing was maintained (see Figure 9.5).

9.7 Conclusion

In situations where all vehicles operate autonomously, the two-tier architecture of the transportation regulation and jamming avoidance is used. We've noticed that traffic signals change dynamically by consuming lengthier sequential sizes and jamming periods. Via detailed simulation results, we confirmed that implementing a 5-tier deep NN system with 400 glia on the concealed strata optimizes our system outcome and provides the finest results. Then we compared the current route's distance and time spent waiting at the intersection to the alternative paths; it redirects automobiles to thoroughfares that take the minimum duration. The outcome indicates an important change in traffic jamming and traffic flow management. Experiments show that our solution is more successful than conventional methods: When the intelligent traffic light module is used, time is reduced by 27 percent, and time is reduced by 18 percent more when the intelligent redirecting phase is used. The outcome reveals the entire output effectiveness of the system up to 41 percent [24,25].

9.8 Acknowledgment

This research work has been supported by RUSA Phase 2.0 Alagappa University.

References

1. W. P. Wagner, "Trends in expert system development: A longitudinal content analysis of over thirty years of expert system case studies," *Expert Systems with Applications*, vol. 76, no. Supplement C, pp. 85–96, 2017.
2. L. A. Zadeh, "Fuzzy logic," *Computer*, vol. 21, no. 4, pp. 83–93, 1988.
3. C. Wu, X. Chen, Y. Ji, F. Liu, S. Ohzahata, T. Yoshinaga, and T. Kato, "Packet size-aware broadcasting in VANETs with Fuzzy Logic and RL-Based parameter adaptation," *IEEE Access*, vol. 3, pp. 2481–2491, 2015.
4. M. Mitchell, *An introduction to Genetic Algorithms*. MIT Press, 1998.
5. B. P. McCune, R. M. Tong, J. S. Dean, and D. G. Shapiro, "RUBRIC: A system for rule-based information retrieval," *IEEE Transactions on Software Engineering*, no. 9, pp. 939–945, 1985.
6. W. S. McCulloch and W. Pitts, "A logical calculus of the ideas immanent in nervous activity," *The Bulletin of Mathematical Biophysics*, vol. 5, no. 4, pp. 115–133, 1943.
7. Z. Chu, D. Zhu, and S. X. Yang, "Observer-based adaptive neural network trajectory tracking control for remotely operated vehicle," *IEEE Transactions on Neural Networks and Learning Systems*, 2016.
8. Y.-C. Yeh and M.-S. Tsai, "Development of a genetic algorithm based electric vehicle charging coordination on distribution networks," in IEEE Congress on Evolutionary Computation (CEC), 2015, pp. 283–290.
9. Y. Yang, J. Hu, and D. Chen, "Research on driving knowledge expert system of distributed vehicle driving simulator," in 11th International Conference on Computer Supported Cooperative Work in Design (CSCWD), 2007, pp. 693–697.
10. J. E. Naranjo, M. A. Sotelo, C. Gonzalez, R. Garcia, and T. De Pedro, "Using fuzzy logic in automated vehicle control," *IEEE Intelligent Systems*, vol. 22, no. 1, pp. 36–45, 2007.
11. J. R. N. Forbes, *Reinforcement learning for autonomous vehicles*. University of California, Berkeley, 2002.
12. J. Z. Zhu, J. X. Cao, and Y. Zhu, "Traffic volume forecasting based on radial basis function neural network with the consideration of traffic flows at the adjacent intersections," *Transportation Research Part C: Emerging Technologies*, vol. 47, pp. 139–154, 2014.
13. M. Kuderer, S. Gulati, and W. Burgard, "Learning driving styles for autonomous vehicles from demonstration," in 2015 IEEE International Conference on Robotics and Automation (ICRA). IEEE, 2015, pp. 2641–2646.
14. S. Shalev-Shwartz, S. Shammah, and A. Shashua, "Safe, multi-agent, reinforcement learning for autonomous driving," arXiv preprint arXiv:1610.03295, 2016.
15. A. E. Sallab, M. Abdou, E. Perot, and S. Yogamani, "Deep reinforcement learning framework for autonomous driving," *Electronic Imaging*, vol. 2017, no. 19, pp. 70–76, 2017.
16. A. Khan, A. Sohail, U. Zahoora, and A. S. Qureshi, "A survey of the recent architectures of deep convolutional neural networks," *Artificial Intelligence Review*, vol. 53, no. 8, pp. 5455–5516, 2020.
17. S.H. Khan, A. Sohail, and A. Khan, "Covid-19 detection in chest x-ray images using a new channel boosted CNN," arXiv preprint arXiv:2012.05073, 2020.

18. F. L. Hall, "Traffic stream characteristics," Traffic Flow Theory. US Federal Highway Administration, vol. 36, 1996.
19. Q. Lin, B. Kwan, and L. Tung "Traffic signal control using fuzzy logic," in *1997 IEEE International Conference on Systems, Man, and Cybernetics. Computational Cybernetics and Simulation*, vol. 2. IEEE, pp. 1644–1649, 1997.
20. T. Brys, T. T. Pham, and M. E. Taylor, "Distributed learning and multi- objectivity in traffic light control," *Connection Science*, vol. 26, no. 1, pp. 65–83, 2014.
21. M. C. Choy, D. Srinivasan, and R. L. Cheu, "Cooperative, hybrid agent architecture for real-time traffic signal control," *IEEE Transactions on Systems, Man, and Cybernetics-Part A: systems and humans*, vol. 33, no. 5, pp. 597–607, 2003.
22. C. Cai, C. K. Wong, and B. G. Heydecker, "Adaptive traffic signal control using approximate dynamic programming," *Transportation Research Part C: Emerging Technologies*, vol. 17, no. 5, pp. 456–474, 2009.
23. P. Mannion, J. Duggan, and E. Howley, "An experimental review of reinforcement learning algorithms for adaptive traffic signal control," in *Autonomic road transport support systems*. Springer, pp. 47–66, 2016.
24. X. Liang, X. Du, G. Wang, and Z. Han, "Deep reinforcement learning for traffic light control in vehicular networks," arXiv preprint arXiv:1803.11115, 2018.
25. S. Maerivoet and B. De Moor, "Traffic flow theory," arXiv preprint physics/0507126, 2005.

10

Deep Learning Solutions for Pest Detection

C. Nandhini and M. Brindha

NIT Tiruchirappalli, Tamil Nadu, India

CONTENTS

DOI: 10.1201/9781003206736-10

10.1 Introduction

In recent years, "object detection" has become one of the most challenging problems to receive attention in computer vision applications. Nowadays, images are available everywhere, and the critical content in an image is the object. Extracting the semantic information from images for further analysis like object recognition and detection is a primary image processing application. There have been numerous endeavors to utilize noteworthy features of object detection for different applications like video surveillance, autonomous driving, human-computer interfaces, robotic vision, crowd counting, anomaly detection, and healthcare. The emergence of deep neural networks (DNNs) leads to learning more complex features effectively. The convolutional neural network (CNN) features have the highest representative power compared with traditional approaches. This article investigates object detection in agriculture; in particular, it focuses on how deep learning-based object detection can be used for automatic pest detection in paddy crops.

10.1.1 Object Detection

Object detection is the primary computer vision task that deals with *what objects are where?* in an image. Most of the vision tasks, in particular, automatic object tracking, instance segmentation, image captioning and anomaly detection, etc., depend on object detection. Object recognition deals with what objects are in an image, while object detection aims to localize the objects in addition to object recognition. Figure 10.1 depicts the schematic representation to differentiate object recognition and object detection. Object recognition identifies the yellow stem borer pest in the image, while object detection identifies the pest and finds the coordinates of the pest location.

Object detection research can be primarily categorized into "general object detection" and "applications using object detection." The goal of general object detection is to devise a technique capable of detecting instances of objects of a particular category, thus simulating human vision. In contrast, the latter utilizes the object detection methods under application scenarios like face detection, pedestrian detection, advanced driver assistance systems, lesion detection, pest detection, etc.

Object detection approaches can be either (i) traditional object detection approach or (ii) deep object detection framework.

- **Traditional object detection:** Traditional object detection methods rely on hand-crafted features. It scans the entire image using a sliding window. It tries to identify the low-level visual cues such as edges, corners, and intensity of the pixels for identifying whether a particular pixel belongs to an object instance. It appears to be simple, but it needs a substantial computational cost. Viola-Jones detectors [1], HOG (histogram of oriented gradients) detectors [2], and DPM (deformable part-based model) detectors [3] are well-known traditional detectors.

- **Deep object detection:** In recent years, deep learning algorithms have attracted the research community leading to various powerful and fast object detection methods. Deep convolutional neural networks (DCNNs) can quickly learn the high-level and feature representations from an image which forms the basis for applying deep learning for object detection.

| Object Recognition | Object Detection |

FIGURE 10.1
Object recognition vs. object detection.

10.1.2 Deep Object Detection

Nowadays, deep learning has outperformed traditional image processing that uses hand-crafted features to recognize and detect objects. The convolutional neural network (CNN) [4–6] based object detectors dominate object detection as it allows the model to learn the complex feature representation automatically from data. Neural networks (NN), with their exceptional capability, infers meaning from incomplete and complex data. NN's are used to extract patterns and spot trends that are difficult for humans to identify. However, the NN for images is computationally expensive since distinct neurons for each pixel might result in thousands of neurons depending on the image resolution.

CNNs [7] are widely used to train 2D images; they process images like a neural network but with reduced neurons. CNN considers the image a matrix, and a filter window will be moved over the matrix for extracting features. Using the same filter throughout the entire image allows for parameter sharing and hence provides translation invariance. Figure 10.2 depicts the architectural design of CNN.

First, the entire image is segmented to produce multiple overlapping sub-regions depending on the size of the convolutional filter. The filter is convolved over those regions one by one to extract the features. Then, training takes place on the feature maps by adjusting the weights of specific neurons. The most significant regions are combined based on the feature maps. The detailed description of various layers is shown in Table 10.1.

10.1.2.1 Types of Deep Object Detection

General object detection frameworks consist of three steps. First, it generates region proposals; a large set of bounding boxes (based on the region of interest) is produced by spanning the entire image. In the next step, the model is trained using the features derived from the proposal regions, which helps in regressing the bounding box coordinates and recognizing the objects. Then, the Intersection over Union (IoU)/Jaccard index measures the area of superimposition between the predicted and ground truth bounding box, which is given as

$$\frac{|gndtruthbox \cap bndbox|}{|gndtruthbox \cup bndbox|} \tag{10.1}$$

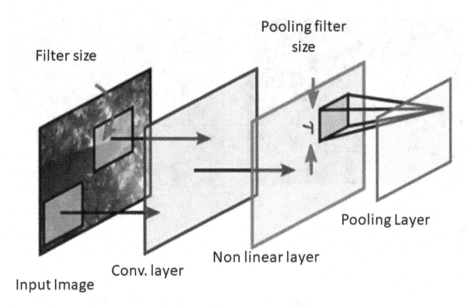

FIGURE 10.2
Convolution neural network (CNN) architecture.

TABLE 10.1

Building Blocks of Convolution Neural Network (CNN) Architecture

Layer	Description
Convolution layer	It contains a set of filters (small square) that convolves with every possible input image location to produce the feature map. It uses spatial locality to reduce the complexity and the number of parameters. More than one filter is used to extract local features such as gradient orientation, edges, color, shape, etc.,
Nonlinear layer	This layer is used to bring in nonlinearity, which helps the network to learn the features by providing normalization via Rectified Linear Unit (ReLU).
Pooling layer	The pooling layer is used to downsample the image by keeping the image content intact by averaging or maximum value from the neighborhood.
Fully connected layer	This layer is used after the image's content is encoded into a feature space where the spatial coordinates are unrelated.

The bounding boxes whose IoU is greater than the particular IoU threshold are chosen for consideration. The higher the IoU threshold, the greater the detection difficulty. If multiple bounding boxes are found with the same IoU, then overlapping boxes are suppressed using non-maximum suppression, which discards the lowest score proposal.

Deep object detection can be broadly classified into two categories based on the steps it takes to detect objects.

Two-stage detector: It detects an object in a two-step process using different networks. As a first step, a limited set of region proposals are generated that may contain the object. During the second stage, they use those proposals for identifying objects with their bounding box coordinates. The two-stage detector like RCNN [4], Fast

RCNN [5], and R-FCN [8] generally makes larger and deeper networks to achieve higher accuracy. However, such good accuracy comes with a high computational expense, which may not be suitable for certain real-world applications.

Single-stage detector: It identifies all the objects in an image using a single-step process; it takes an input image and learns the bounding box coordinates with object class probabilities using a simple network. The single-stage detector such as Single Shot Detector (SSD) [6], RetinaNet [9], and YOLOv3[10] are faster than two-stage detectors with lower accuracy rates. Most single-stage detectors use anchor boxes as an alternative to sliding window proposals to refine the prediction of bounding boxes. Anchor boxes are a predetermined group of boxes with different dimensions and aspect ratios depending on the size of the objects in the dataset. During training, with each pixel as the center, multiple bounding boxes depending on the anchor box sizes are generated as proposals and matched to the object instance according to IoU overlap. These anchor-based detectors lack more generalization ability and also involve complex ground truth matching using Jaccard Index. The anchor-free detectors such as CornerNet [11], CenterNet [12], and Fully Convolutional One-Stage Object Detection (FCOS) [13] overcome the drawbacks of anchor-based approaches and achieve a higher average precision (AP) with faster inference.

10.1.3 Challenges in Object Detection

The main challenges of generic object detection are recognizing and localizing the objects in images with high accuracy and efficiency. Accuracy and efficiency are both competing goals. The detection should meet the real-time speed with less memory and high efficiency. Imaging conditions, weather, pose variation, shade, occlusion, noise, distortions, and object deformation also affect the quality of object detection. The common challenges are shown in Figure 10.3.

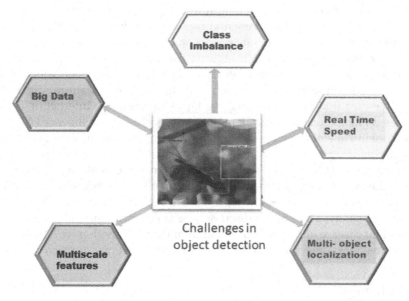

FIGURE 10.3
Challenges in object detection.

Big data: Creating large-scale annotated datasets is a tedious task; it involves collecting a diverse set of images for each target object category. Annotating it manually using crowdsourcing is labor-intensive. Augmenting images in the dataset is carried out by perturbing an image by adding noise, flipping, rotating, scaling, etc.

Real-time speed: In recent trends, object detection algorithms try to balance speed and accuracy. Single-shot detectors significantly reduce the computation time; however, the detection accuracy drops off. Based on the design choice, the object detection model can improve only accuracy or speed.

Multiscale features: The detector should detect the objects efficiently as it may appear in varying sizes in images. Anchor boxes, feature maps from multiple layers, and feature pyramid networks are methods to obtain multiscale features.

Multi-object localization: Human vision to recognize numerous object categories is impossible using computer vision. The existing datasets contain a limited number of object categories; COCO dataset with 80 objects, Pascal VOC with 20 objects, ImageNet with 1,000 object categories, etc.

Class imbalance: The backgrounds, which are different from objects, is considered negative samples, and it is comparatively higher than the objects. Hence, during training, this creates a class imbalance.

10.2 Advances in Agriculture

Humanity has practiced agriculture for centuries, and it is a labor-intensive job requiring more physical strength. Farming is the primary source of income for many rural people and contributes to the national economy through processed food and commodity exports. The quantity and quality of crop yields are determined by extrinsic variables such as temperature, climatic conditions, and environmental factors. With limited farmland, lack of fresh water, and continually changing climate conditions, food production must be increased to feed an ever-increasing population. In traditional farming, the same crop-cultivation methods are used regardless of the nature of the land. Fertilizers and insecticides are used irrespective of diseased or nutrient crops, which leads to wastage of manure. "Smart farming" is an innovative idea that refers to managing farms utilizing the Internet of Things (IoT), robots, drones, a global positioning system (GPS), and artificial intelligence (AI) to raise the quantity and quality of yields while reducing the amount of manual labor. The rapid growth of low-cost digital imaging instruments leads to advanced plant phenotyping techniques that attract the research community for effective crop management. Each farm is examined to determine the suitable crops and water requirements for maximum production. Early identification and application of pesticides to the affected area alone reduce pesticide usage and increase food safety. This section discusses the smart farm and the application of deep learning for plant disease and pest identification in detail. This chapter's final section also introduces a novel pest detection method, using a deep learning approach.

10.2.1 Smart Farming

The main motive of smart farming is to increase crop quality and quantity while reducing human effort. Smart farming is a cyclical data gathering method used for crop management and decision-making, with the cycle continuing in each consecutive year. Data are

TABLE 10.2

Technologies Used in Smart Farming

Technologies	Usage
Sensors	Used in monitoring the temperature, moisture content in the soil, water, and light conditions
Location system	GPS, Satellite
Robots, drones	Helping on the farm with plowing, planting, harvesting, spraying pesticides, and crop monitoring
Communication system	A mobile connection is used to connect over the internet
Software applications	IoT-based specialized application for specific farm types
Precision irrigation	Used to give the right amount of water solely to the plant's root, only when needed
Cloud technology	Storing and analyzing the big data received from the field
Deep learning solutions	To help in making decisions and predictions based on the big data (diagnose the plant disease, pest detection)

kept in a database (library) each year and utilized as historical data for future decision-making. Smart farming is crucial for tackling crop disease, sustainability, food security, and environmental impact issues in agricultural production. Remote sensing is used on a larger scale to predict weather and earth properties using unmanned aerial vehicles (UAVs) and satellites. Small-scale crop management in smart farming is accomplished by using sensors, cameras, and mobile phones that collect data from the agriculture ecosystem regularly. The data is transmitted over the internet using IoT and then saved on a cloud-based platform for further processing. The collected data are analyzed based on the specific requirements of the crop using deep learning with little human participation. As a result, farmers can respond quickly to newly arising concerns and changes in the environmental condition. The technologies used in the agricultural land for smart farming are given in Table 10.2.

10.2.2 Deep Learning in Agriculture

Deep learning is an extension of neural network architecture with more layers for processing than the traditional neural network. The advances in computational systems, particularly graphical processing units (GPU), hardware, and analysis of massive amounts of data, have made deep learning models computationally possible. The deep learning techniques make smart agriculture possible with various applications in agriculture, including leaf classification, plant disease, pest identification, yield estimate, weed detection, soil moisture/nutrient prediction, and weather forecasting. Drones acquire multispectral, thermal, and visual imaging while hovering. The data collected by them provides extensive metrics that help farmers to monitor plant health, field water pond mapping, chlorophyll quantification, drainage designing, and so on.

Most of the data acquired from agricultural farms using cameras or UAVs for smart farming are in the form of images. Image analysis using image processing (HOG, SIFT features) techniques, machine learning (SVM, k-means, and wavelet-based filtering) based approach, and DL are typical for image classification and detection tasks. The advancement of deep learning has significant potential in applications domains like healthcare, agriculture, driver assistance system, etc. The role of deep learning in agriculture has been explored for tasks such as land cover classification, plant recognition, crop type classification, identification of weeds, plant disease, pest detection, crop yield estimation, fruit counting, etc.

10.2.3 Automatic Pest Detection

Pests in plants have caused significant economic, social, and environmental harm since the emergence of agriculture. Systematic monitoring and early diagnosis are required to take preventative steps at an early stage. Farmers have typically recognized illnesses based on visual cues and experience. Advances in analyzing enormous amounts of data and computer vision provide the opportunity to adopt precision farming practices and expand the market for computer vision applications in smart agriculture. The key motivation for using the deep learning-based plant illness/pest detection exemplar is to provide the farmers with an easy-to-use innovative tool that detects early-stage infections with its preventive measures, thereby increasing crop cultivation and improving the happiness of paddy farmers.

10.2.4 Challenges in Automatic Pest Detection

One of the most fundamental and vital operations in agriculture is detecting the pest in a crop. Pest identification involves a high degree of complication due to the visual examination of symptoms on plant leaves. Some of the common challenges which may lead to inaccurate diagnosis of specific pests by expert agronomists and plant pathologists are

- Various stages of pests (such as worms and insects) may vary.
- Due to atmospheric and soil conditions, the appearance of pests may vary.
- Pests can arise in almost every part of the plant, such as leaf, trunk, grain etc.
- Most of the pests are relatively small compared to other generic object detection. Hence the detector for pest detection should be robust enough to detect small pests.

The availability of an automated computational system to identify the pest in a reliable and fastest way would greatly benefit the farmers. The intrinsic and extrinsic factors affecting pest detection are discussed in this section. Figure 10.4 shows sample images from the IP-102 dataset with the factors affecting pest detection.

10.2.4.1 Extrinsic Factors

Image background: Although the images acquired for training may have a clear background, in real time, leaves with crowded environments of soil, plants, and leaves may interfere with the precise diagnosis of pests. Deep learning models can learn things in the presence of a noisy background. However, specific backgrounds are similar to those of the leaves with a particular type of pest, inhibiting the network learning capability and leading to errors.

Image capture condition: The device used for image capture has an impact on the image's quality. Images can be acquired from different equipment by various people under variable situations, such as lighting, timing, shadow, camera angle, and picture resolution, all of which can significantly affect the image's qualities.

(a) Life-cycle of pest   (b) Angle variation, multiple pests, background

FIGURE 10.4
Factors affecting pest detection.

Furthermore, while the photographs are transferred over the internet, the information quality may degrade owing to compression, resulting in errors in meaningful analysis.

10.2.4.2 Intrinsic Factors

Similar appearance: A few pest/plant diseases look identical and have similar symptoms that are difficult to diagnose merely by looking at the image. To resolve the ambiguity in identifying the pests, further information such as prevailing climatic conditions and the disease history of that plant may be required. Images captured in the infrared range can also help resolve the issue, although at a higher expense.

Environmental factors: Most diseases cause variation in color, shape, and texture in the affected regions. Environmental elements such as temperature, wind, and humidity can also unpredictably introduce variation. In the early stages, the leaves may be very faint; hence, the pest is unnoticeable, leading to a misdiagnosis. It is critical to have images of all stages of the pest, with a wider variety of variances will be considered for robust performance.

Multiple pests: As the plant's immune system deteriorates, a single plant may have various pests. Some pests are too small and are in a group, so annotating them is difficult. However, this increases the number of samples, and having less data for this type of image may cause data imbalance.

10.2.4.3 Big Data Availability for Deep Detection

Creating an extensive dataset for pest detection is a time-consuming and tedious task. Deep learning requires a massive number of images for a specific type of pest.

Image capturing: Image capturing by visiting multiple fields during various stages of pest and taking images of the healthy and affected region is one of the significant and laborious works for creating the dataset.

Annotating the dataset: To annotate the image, an expert's help or a laboratory test may be required to identify the pest and its severity. As the labeling is based only on visual cues, sometimes it may also affect the reliability of the dataset. Some of the pests in the images are too small and hence difficult to annotate.

Less data for rare species: There may be unavailability of a reasonable amount of data for some rare species. So, common pests are easily identified, while rare species are more likely to be misclassified.

Domain adaptation: Testing images from a dataset on which the model is trained gives higher accuracy, while the cross-dataset evaluation lowers the performance significantly. To overcome these, images should be collected from different geographical regions to include various species with other climatic and imaging conditions.

10.3 Novel Smart Intelligent System for Paddy Pest Detection

Figure 10.5 shows the components of the paddy pest detection system, using a deep learning approach. The system is divided into two phases:

Training phase: This involves collecting images, training with the object detection model and deploying the trained weights in the server. During this phase, an extensive collection of paddy crop images with various pests should be collected

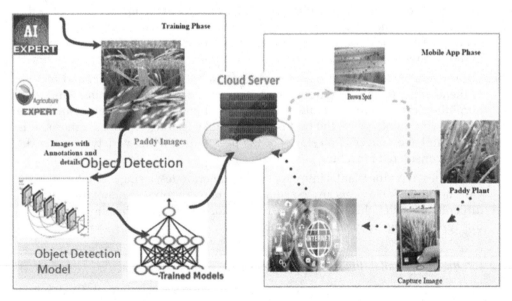

FIGURE 10.5
Components of the proposed system.

from the agriculture field. Then the images are annotated with the type of pest, age of the plant in days, the stage of the disease, parts affected, and suggested remedial steps with the help of both agriculture and AI experts. Then, the image dataset is trained and fine-tuned using the transfer learning-based EfficientDet model. The trained model should be deployed in the cloud server.

Mobile app phase: With the help of a portable mobile device, the farmer captures an image of the affected paddy crop. Then, the mobile sends the image to the server connected with the network. The server detects the type of pest in the paddy crop with its severity along with one or more remedial steps.

10.3.1 Related Work

Damage to rice crop occurs during production due to natural calamities such as rainfall, flood, and several disease and pests affecting the rice crop. So, spotting the pests in the paddy crop is an essential task for the farmers, so that protective measures can be taken at the earliest stage. Nowadays, deep learning techniques have made progress in various image classification tasks. The application of deep learning in agriculture is mainly used to identify plant diseases and pest which causes severe damage to the crops, thus affecting the economy. Recent literature [11–15] shows that deep learning techniques are used to analyze diseases in rice, wheat, maize, tomato, apple, tea, pear, peach, grapevine, etc. Karmokar et al. [11] use a neural network ensemble (NNE) to recognize the disease in leaves of the tea plant. Mohanty et al. [15] explore deep CNN-based disease detection for 15 varieties of the crop with a total of 26 diseases. The authors used a dataset collected from different versions of the PlantVillage dataset. Sladojevic et al.[14] suggested a system that identifies healthy leaves and 13 different plant disease types using the images from the internet. With 3,000 images from the dataset, after applying appropriate transformations, they can train with a dataset of approximately 30,000 images. Fused the texture and shape features in a CNN to detect quick decline syndrome on olive leaves with the help of a limited dataset. Wang et al. [17] estimate the level of damage caused by the plant disease using a simplistic network that uses fine-tuning strategy and compares that with the training from scratch approach. Atole and Park [18] used AlexNet-based architecture to classify the rice plant as a diseased, normal, and snail-infested crop. Rahman et al. [19] proposed a two-stage lightweight CNN model to identify eight types of rice plant disease suitable for mobile devices.

Zhou et al. [20] used a full convolution network-based detector to count the number of panicles per unit area of the field to estimate the crop yield. Barrero and Perdomo [16] fuse RGB image and multispectral image features to spot the Gramineae weed in the rice fields. Sourav Kumar Bhoi et al. [21] proposed a rice pest detection model using an unmanned aerial vehicle (UAV) controlled using the Internet of Things (IoT). The previous works in the literature mainly aim for accurate classification or detection of disease in agriculture crops. Several CNN architectures like AlexNet, VGG16, and GoogLeNet are explored for this purpose.

Hassan et al. [24] proposed an automated system that uses color and shape-based feature descriptors to recognize grasshoppers and butterflies in colored images. Use a combination of SIFT and local features like contour, geometric, and texture features to recognize insects. The classifiers such as linear classifier, decision tree, and k-nearest neighbor are explored as part of their work.

The state-of-the-art plant disease classification has achieved remarkable success. However, very little literature focuses on detecting pests in the rice crop. Moreover, to be used by the mobile application, the number of hyper parameters should be reduced so that it is suitable for real-world mobile application deployment.

Though this deep learning-based object detection performs well, the main limitation of these architectures is the number of parameters to train. However, farmers in remote regions of developing countries have mobile devices with limited memory and slow-speed internet connectivity. Hence, a mobile application capable of connecting to the cloud with the trained pest detection model is proposed in the system.

In summary, the proposed article makes three significant contributions for identifying rice disease and pest detection. First, this chapter focuses on a preliminary study of detecting five different pests that cause damage to the rice crop. Second, the state-of-the-art EfficientDet [22] detector has been investigated for pest detection using the images collected from agricultural environments. A mobile device-based application is proposed, which can be used as an effective tool for farmers.

10.3.2 Training Phase

The training phase involves collecting images and annotating them. IP 102 Pest classification dataset is used for that purpose. The proposed model uses the pre-trained weights from EfficientDet [23] model trained on the COCO dataset for the pest detection task.

To achieve this, a workflow of the proposed work is illustrated in Figure 10.6. The first module takes the images from IP102 for rice disease pest detection as input. Next, the images are annotated to localize the pest with their name; the cropped images with their annotations from the rice disease pest detection images dataset. In the third step, the EfficientDet model is used for fine-tuning the proposed pest detection system. Once the model is trained with the custom dataset, the trained weights can be used to detect and classify the pests on the test set.

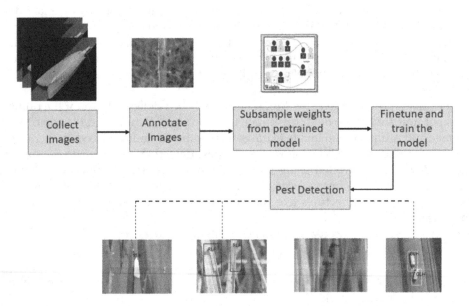

FIGURE 10.6
Workflow of the proposed system.

10.3.2.1 Dataset Description

In the course of the proposed work, IP102 classification dataset for pest recognition is used. It contains 102 categories of pest that affects the agricultural product. The IP102 dataset is one of the large-scale datasets for pest recognition with 75,000 images. The images under this dataset have a hierarchical structure; the top level is based on economic crop or field crop. Paddy, wheat, and rice belong to field crops, while lemon belongs to economic crops. Next level, the pests affecting a particular crop are grouped together. There are 14 different classes of insects that affect the rice crop available in the dataset.

In pest detection, the number of images for the particular insects in the datasets is insufficient. The lack of samples is a challenge to the deep learning detection methods. Therefore, augmentations are done on the IP102 dataset to enhance the training effect and model robustness. The training data is augmented by employing some of the equivalent transformations to the images along with the bounding boxes for pest detection. Some conventional types of transformation like random cropping and arbitrary rotation are used to enlarge the dataset.

However, for training the model, annotated images are needed. As no public dataset is available, the images of particular pests collected from the IP102 classification dataset are annotated using the open-source labeling tool as it is easy to install and user-friendly. A bounding box is drawn to localize each pest and labeled based on the category. After localizing the image, the rectangular bounding box coordinates and category information are saved in the generated XML file using the standard VOC2007 format.

The issues encountered in the real-world dataset are occlusion, overcrowded images, and incomplete annotations. The pest with shadows, varying backgrounds, different directions, angles and occlusion shows the diversity of the custom dataset to give robust detection

10.3.2.2 Classes Used in the Proposed Method

The rice crop is likely to be infected by more than 100 insect species at any one of the stages. The major ones, which cause a severe threat to rice production, are brown plant hopper (BPH), rice leaf folder/roller (RLF), green leaf hopper (GLH), white-backed plant hopper (WPH), and yellow stem borer (YSB). So, the proposed system is trained to detect those five pests of major economic significance. Figure 10.7 depicts the classes used in the proposed work with the examples from the custom dataset.

The proposed work aims to detect pests in the rice crop. Toward this goal, the model is trained to detect five classes, as shown in Table 10.3. The model is trained to learn the dissimilarity between the pests so that a robust performance is expected from the detector.

10.3.3 EfficientDet

EfficientDet improves performance and efficiency by introducing several vital optimizations to the existing object detectors. Any object detectors generally have three main blocks: backbone network is used to draw meaningful representation from the given image; the second block is a feature network that fuses multiple levels of features from the backbone and outputs the salient characteristics of the image; the third block is classification and localization head to predict the instance class and bounding box coordinates. Figure 10.8 shows the overall architecture of EfficientDet. One of the key optimizations was the introduction of the EfficientNet backbone with compound scaling and neural architecture

Yellow Stem Borer   Green Leaf Hopper

Leaf Folder   White-backed Plant   Brown Plant Hopper
                     Hopper

FIGURE 10.7
Classes of pest.

TABLE 10.3

Major Pests and their Significance

Pest	Appearance	Causes	Significance
RLF	The caterpillars of this type of pest fold the rice leaf around themselves with silk strands.	High humidity, shady areas, rainy season, heavy use of fertilizer, surrounded by grassy weeds	Can cause yield loss, ecological disruptions
YSB	Triangular and medium-sized butterfly, with yellow color bugs.	Rainfall and humidity	Destroy crops at any stage
BPH	Crescent shape white eggs and brown nymphs.	Rainfall, wetlands, high shade and humidity	Drying out of the plants, hopper burns
GLH	Greenish transparent eggs, wedge-shaped bright green color insects with variable black markings.	Staggered planting, rainfall, and wetland	Transmit viral diseases such as tungro, yellow dwarf, etc.
WPH	Cylindrical eggs and forewings are uniformly hyaline with dark veins.	Close and staggered planting	Leaves turn orange-yellow from tip to mid vein, reducing grain production

search. EfficientNet backbone is fundamentally a sequence of convolution (An Inverted Residual Block (MBConv)) blocks and the compound scaling method helps in scaling up multiple dimensions like network width, height, and image size simultaneously, resulting in performance increment. The scaling factor realized to scale the EfficientNet were:

$$\alpha = 1.2, \beta = 1.1 \text{ and } \gamma = 1.15 \tag{10.2}$$

where α is the depth factor, β – the width factor, and γ the resolution factor.

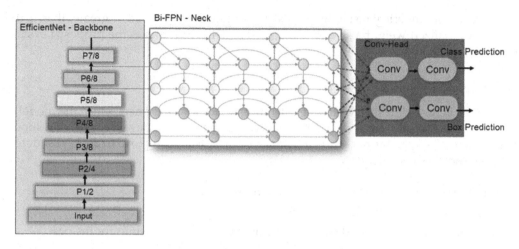

FIGURE 10.8
EfficientDet architecture (adapted from *Tan, M., Pang, R., & Le, Q. V. 2020.*).

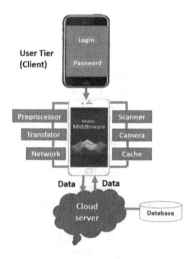

FIGURE 10.9
Mobile service framework.

The optimized parameters, as in Equation (10.2), are used in the training phase for rice crop pest detection. The EfficientNet pre-trained on ImageNet is used for the proposed work.

Another optimization done in feature networks is the introduction of fast normalized multilevel feature fusion, which allows information flow in both upward and downward directions. The features extracted from EfficientNet are used with repeated layers of Weighted Bi-directional Feature Pyramid Network (BiFPN). The BiFPN layer fuses multi-scale features in a bi-directional path assigning weights to each feature. These feature weights are learned using the Normalized Fast Fusion technique, as shown in Equation (10.3), which is simpler and significantly improves efficiency.

$$O = \sum_i \frac{W_i}{\varepsilon + \sum_j Wj} . I_i \qquad (10.3)$$

where to avoid numerical instability $\varepsilon = 0.0001$ is used and W_i should be greater than 0; the output normalized weight also falls between 0 and 1 which makes the proposed system much more efficient.

10.3.4 Server Framework

The server is hosted in the cloud and provides service to the farmers. Once the model is trained using EfficientDet, it should be deployed in the server. The server should respond to the client when the farmer sends the image of the affected crop. Apart from pest detection, the server is responsible for

- Maintaining authorization information of the user
- Allow access to an authorized user
- Maintaining the knowledge base comprises preventive measures for the pest
- Provides the reason for the cause of pest and control measures

10.3.5 Mobile Service Framework

The suggested solution is a mobile service framework, a three-tier architecture for diagnosing pests on rice crops. The client component encompasses a set of user interfaces that allows farmers to register and login. Then, authorized farmworkers can transfer the image of an unhealthy plant focusing on the affected region with pests (if any) to the server, which in turn identifies the disease. To provide all-time availability, the server – a standard web server – is hosted in the cloud and accessible over the internet. The database resides in the server and contains the collection of disease types, and remedial steps to alleviate the disease. Furthermore, it also maintains the preventive measures to be taken to avoid the pathogens in future agricultural cycles. The best procedure to apply manure and chemical pesticides for controlling the pest/disease is also given by the server. The client application also contains an off-line repository of disease/pests together with mitigative measures that can be used by the farmers without network connectivity. The central component resides on the mobile phone, which plays the intermediary role and functions as the user interface. It is used to capture the diseased plant's image, preprocess it, and route it to the server. It also caches the response received from the server to help the farmers for off-line usage.

The functions of the client include image acquisition and providing a knowledge base of disease prevention and control. The function of the server includes image processing, feature extraction and identification of the pest. The system's recognition process goes as follows: when pests' images were collected in fields and sent to the server by an android smartphone, spotting the pest was carried out by the server.

10.3.6 Performance Metrics for Evaluation

This section provides insights into performance metrics that are commonly adopted for evaluating the performance of an object detection model.

10.3.6.1 Precision and Recall

The object detection framework generates very imbalanced data. In object detection, there will be a more significant number of background classes than the detectable classes.

Precision-recall is one of the metrics used in such cases. For matching the ground truth box with multiple default boxes, IoU overlap is used as the metric. The IoU threshold value can be set to determine whether or not the detected object is valid.

- *IoU is 1:* This case occurs when the ground truth bounding boxes perfectly overlap with the predicted box.
- *IoU > threshold:* The superimposed area covered by the predicted and original box is larger than the IoU threshold, hence considered valid and classified as True Positive (TP).
- *IoU < threshold:* The IoU between the actual and predicted bounding box is lesser than the threshold value, this is considered invalid and classified as False Positive (FP).
- There are actual boxes in the image, but there is no predicted box. This case comes under False Negative (FN).

Precision denotes the percentage of correct predictions made by the algorithm out of total predictions.

$$\text{Precision} = \frac{TP}{TP + FP} \tag{10.4}$$

The recall ratio represents the percentage of correct predictions out of the ground truths detected by the algorithm.

$$\text{Recall} = \frac{TP}{TP + FN} \tag{10.5}$$

10.3.6.2 Average Precision (AP)

The model's accuracy is gauged by the average precision (AP), which is estimated using the precision-recall curves. The comparison between various object detection models can be done using AP, as it gives the numerical metrics, making the comparison easier.

AP is computed effectively by interpolating the precision through all k unique recall r values, as proposed in Pascal VOC 2010 Challenge as follows

$$\text{Average Precision} = \sum_{k=0} P_i(r_k) \tag{10.6}$$

where $P_i(r_k) = \max(p\left(\hat{k}\right)$

where $P_i(r_k)$ is the interpolated precision that takes the maximum precision over all recalls greater than k.

10.3.6.3 Mean average Precision (mAP)

The average AP value of all (C) classes represents the mean average precision and is calculated using the formula as follows

$$mAP = \frac{\sum_{i=1}^{C} AP(i)}{C} \qquad (10.7)$$

The weighted average of AP is used when there is a difference in the number of samples in each class. For all C classes, weights w_i are assigned, based on the number of class samples available in the dataset, and all the weights will sum to one.

$$weighted\, mAP = \frac{\sum_{i=1}^{C} W_i AP(i)}{C} \qquad (10.8)$$

10.3.6.4 Precision-Recall Curve

One of the useful measures for the object detection framework is the precision-recall curve. The precision-recall curve helps identify the best threshold value for detection by plotting precision and recall for different threshold values. The area covered by the precision-recall curve is another performance indicator. The higher the area covered, the higher the precision and recall.

10.3.6.5 Inference Speed

The inference speed is an essential measure where fast detections are required. To be used in a real-time environment such as surveillance applications, inference speed is one of the standard performance metrics. Frames per second (FPS) is the standard measure to evaluate the inference speed. It calculates how many pictures can be processed by the algorithm in a second. To give good visual effects, 25 FPS has to be achieved.

10.3.6.6 Service Time for User

After taking the image using the mobile app, the user has to send the image to the server deployed in the cloud. Let t_M be the image transmission time to the server, and the processing time at the server be t_P. The pest information needs to be sent to the user with a transmission time of t_C. So the total service time encountered by the user is given by Eqn (10.9), where t_I is the propagation delay introduced by the internet in both the direction

$$Service\ Time = t_M + t_P + t_C + 2t_I \qquad (10.9)$$

10.4 Conclusion

This chapter has introduced object detection, smart farming, and deep learning for the agricultural domain. The suggested cloud-based pest detection framework can act as an immutable architecture in the agriculture domain where all farmers can efficiently access the pest detection service. Thus, it helps in the decision-making process for selecting the particular chemical pesticides regardless of their experience level. To ensure efficient detection, an EfficientDet-based network is proposed for pest detection. The article also investigates various performance metrics for the efficacy of the framework for object detection. This prototype guarantees the availability of pest detection services over the cloud to help improve crop quality and yields, thus supporting sustainable development.

References

1. P. Viola & M. J. Jones (2004). "Robust real-time face detection," *Int. J. Comput. Vis.*, vol. 57, no. 2, pp. 137–154.
2. N. Dalal & B. Triggs, "Histograms of oriented gradients for human detection," in *Computer Vision and Pattern Recognition, 2005. CVP'R 2005.* IEEE Computer Society Conference on, vol. 1. IEEE, 2005, pp. 886–893.
3. P. F. Felzenszwalb et al. (Sep. 2010). "Object detection with discriminatively trained part-based models," *IEEE Transactions on Pattern Analysis and Machine Intelligence*, vol. 32, no. 9, pp. 1627–1645.
4. R. Girshick, J. Donahue, T. Darrell, & J. Malik (2016). "Region-based convolutional networks for accurate object detection and segmentation," *IEEE Transactions on Pattern Analysis and Machine Intelligence*, vol. 38, no. 1, pp. 142–158.
5. S. Ren, K. He, R. Girshick, & J. Sun (2015). "Faster r-cnn: towards real-time object detection with region proposal networks," *Advances in Neural Information Processing Systems*, vol. 28, pp. 1–9.
6. S. Zhang, L. Wen, X. Bian, Z. Lei, & S. Z. Li, "Singleshot refinement neural network for object detection," in *IEEE CVPR*, 2018.
7. A. Krizhevsky, I. Sutskever, & G. E. Hinton, "Imagenet classification with deep convolutional neural networks," in *Advances in neural information processing systems*, 2012, pp. 1097–1105.
8. J. Dai, Y. Li, K. He, & J. Sun, "R-fcn: Object detection via region-based fully convolutional networks," in *Advances in neural information processing systems*, 2016, pp. 379–387
9. T.-Y. Lin, P. Goyal, R. Girshick, K. He, & P. Dollar, "Focal loss for dense object detection," in *IEEE Transactions on Pattern Analysis and Machine Intelligence*, 2018.
10. J. Redmon, & A. Farhadi (2018). "Yolov3: An incremental improvement," arXiv preprint arXiv:1804.02767.
11. B. C. Karmokar, M. S. Ullah, M. K. Siddiquee, & K. M. R. Alam (2015). "Tea leaf diseases recognition using neural network ensemble," *International Journal of Computer Applications*, vol. 114, no. 17. doi: 10.5120/20071-1993
12. A. Fuentes, S. Yoon, S. C. Kim, & D. S. Park (2017). "A robust deep-learning-based detector for real-time tomato plant diseases and pests recognition," *Sensors*, vol. 17, no. 9, 2022. doi: 10.3390/s17092022
13. G. Wang, Y. Sun, & J. Wang (2017). "Automatic image-based plant disease severity estimation using deep learning," *Computational Intelligence and Neuroscience*, 2017. doi: 10.1155/2017/2917536

14. S. Sladojevic, M. Arsenovic, A. Anderla, D. Culibrk, & D. Stefanovic (2016). "Deep neural networks based recognition of plant diseases by leaf image classification," *Computational Intelligence and Neuroscience*, vol. 2016, pp. 1–11.

15. S. P. Mohanty, D. P. Hughes, & M. Salathé (2016). "Using deep learning for image-based plant disease detection," *Frontiers in Plant Science*, vol. 7, p. 1419.

16. O. Barrero, & S. A. Perdomo (2018). "RGB and multispectral UAV image fusion for Gramineae weed detection in rice fields," *Precision Agriculture*, vol. 19 no. 5, pp. 809–822.

17. G. Wang, Y. Sun, & J. Wang (2017). "Automatic image-based plant disease severity estimation using deep learning," *Computational Intelligence and Neuroscience*, vol. 2017, pp. 1–8.

18. R. R. Atole, & D. Park (2018). "A multiclass deep convolutional neural network classifier for detection of common rice plant anomalies," *International Journal of Advanced Computer Science and Applications*, vol. 9, no. 1, pp. 67–70.

19. C. R. Rahman, P. S. Arko, M. E. Ali, M. A. I. Khan, S. H. Apon, F. Nowrin, & A. Wasif (2020). "Identification and recognition of rice diseases and pests using convolutional neural networks," *Biosystems Engineering*, vol. 194, pp. 112–120.

20. C. Zhou, H. Ye, J. Hu, X. Shi, S. Hua, J. Yue, … & G. Yang (2019). "Automated counting of rice panicle by applying deep learning model to images from unmanned aerial vehicle platform," *Sensors*, vol. 19, no. 14, p. 3106.

21. S. K. Bhoi, K. K. Jena, S. K. Panda, H. V. Long, R. Kumar, P. Subbulakshmi, & H. B. Jebreen (2021). "An Internet of Things assisted Unmanned Aerial Vehicle based artificial intelligence model for rice pest detection," *Microprocessors and Microsystems*, vol. 80, p. 103607.

22. A. C. Cruz, A. Luvisi, L. De Bellis, & Y. Ampatzidis (2017). "X-FIDO: An effective application for detecting olive quick decline syndrome with deep learning and data fusion," *Frontiers in Plant Science*, vol. 8, p. 1741.

23. M. Tan, R. Pang, & Q. V. Le, Efficientdet: Scalable and efficient object detection. in *Proceedings of the IEEE/CVF conference on computer vision and pattern recognition*, 2020. pp. 10781–10790.

24. Hassan, S. N., Rahman, N. S., & Win, Z. Z. H. S. L. (2014). "Automatic classification of insects using color-based and shape-based descriptors," *International Journal of Applied Control, Electrical and Electronics Engineering*, vol. 2 no. 2, pp. 23–35.

11

Deep Learning Solutions for Pest Identification in Agriculture

Monika Vyas and Amit Kumar

Indian Institute of Information Technology Kota, Rajasthan, India

Vivek Sharma

Malviya Nagar Institute of Technology Jaipur, Rajasthan, India

CONTENTS

11.1 Introduction

The main objective of agricultural activities throughout human history has always been to make them more economically efficient. However, several challenges regarding quality and cost have not been addressed. For better agricultural products, we should visit agricultural production areas frequently, Hence can take all the necessary precautions during crop production. We seek to identify that agriculturists waste a lot of time and resources on

DOI: 10.1201/9781003206736-11

several farm visits, and after that, crop production is increased. So precision agriculture is essential because agriculturists spend a lot of time evaluating and monitoring their crops. Due to the advancement of technologies through IoT, proper monitoring can be done that enables crop management to be efficient, and labor costs are decreased. However, the before-mentioned technique is not enough to create such a smart farming environment. During the process of smart farming, observation, diagnosis, decision, and action should take place. A continuous data collection and processing loop is required to make rapid improvements to optimize the agricultural process. Moreover, the data can be recorded and collected with the help of several sensors to capture the various features of soil, crop, biodiversity, and livestock. Machine learning techniques are a very useful component for deciding the various actions. Finally, the end-users evaluate the conditions and increment the rule. A passion for agriculture is insufficient to be a farmer in today's world. Farmer should have expertise in accounting, agriculture, economics, law, and data analysis to achieve precision agriculture. Unfortunately, a high level of expertise is unrealistic in some regions because most agricultural enterprises are family farms. Pesticides and fertilization can be used to establish farming techniques leading to several effects on the real-world environment. After developing consciousness, it became clear that instead of treating all plants the same, each plant should be treated differently depending on its needs. However, IoT and machine learning are key components of the intelligent agriculture system. This type of advice is available at an affordable price for farmers, and with these systems, the latest technologies are used to automate crop monitoring, requiring minimal human interaction.

11.2 Existing Literature

The highest number of papers dealing with agriculture-relevant deep learning is present on the SCI database, and none existed before 2016. In Table 11.1 we offer a time trend analysis. This table depicts the top eight most productive countries in terms of articles in 2019. Throughout this period, China was a leader. Identically, the United States increased its publications at a much faster rate than the other six large countries. Except for publications

TABLE 11.1

Countries with Highest Productivity

Country	2019	2018	2017	2016
China	30	5	3	2
USA	9	3	2	—
Spain	4	2	—	—
France	2	3	—	—
Australia	3	2	—	—
Turkey	4	1	—	—
Denmark	3	—	1	1
Italy	4	1	—	—
Others	17	10	6	2

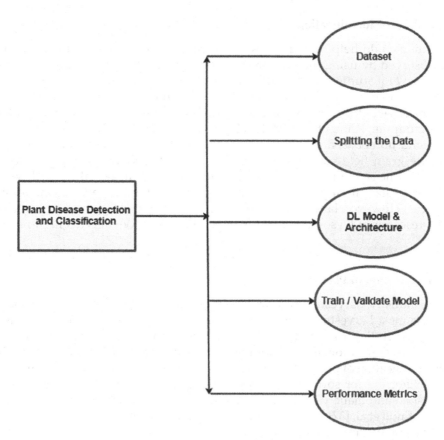

FIGURE 11.1
Deep learning process.

by Chinese writers which focus on topics, the ratio of topics focused on in all the countries is approximately equal. Detection of disease, recognition of objects, precision livestock farming, recognition of pests, and land cover identification are among the topics covered.

11.2.1 Disease Detection

Plant pests are one of the most common causes of agriculture productivity losses. It is necessary to monitor the product's condition and prevent diseases from propagating. The methods for preventing and diagnosing plant diseases vary from one plant to another. The techniques for recognizing plant-specific diseases have been described in the literature. Kerkech et al. [20] developed deep learning techniques for identifying vine diseases that were recognized by applying colorimetric spaces and then implemented to UAV images. Lu et al. [21] suggested an approach for detecting wheat diseases that utilizes a learning-based sensor to detect illness in tomatoes. In Hu et al. [22], a low shot learning method was proposed to detect disease in tea leaves. Coulibaly et al. [23] suggested a technique for identifying mildew diseases in pearl millet, using a transfer learning approach and feature extraction. Cruz et al. [24] suggested artificial intelligence-based technique for identifying symptoms of yellows grapevine.

11.2.2 Land Cover Identification

Deep learning could help find land cover and cover type. Plant growth stages are frequently observed by using multi-source satellite images. However, deep learning was applied in several studies to evaluate land production and classify land cover. Kussul et al. [26] proposed a procedure for using high-resolution satellite data to develop sustainable objective indicators. To determine field borders and utilize a full CNN globally and with grouping, Zhou et al. [28] developed a classifier based on deep learning that classifies land units based on time series features of crops. Yang et al. [29] suggested that the rice grain field can be identified by a deep CNN. This technique can make estimations at the ripening stage by using remote sensing images taken by UAVs. Zhao et al.[30] suggested an approach for rice layout that integrates a CNN model with a decision tree. However, land cover classification does not entirely depend on satellite data. The research of Dyson et al. [31] is dependent on UAV-collected high-resolution images.

11.2.3 Classification of Plants

In the fruit industry, harvesting consumes a lot of time and is a labor-intensive operation, and because most harvesting is performed manually, rapid development is focused on automating harvesting robots. Researchers have been focusing on crop-specific automation systems because automation techniques cannot be generalized. Grinblat et al.[33] propose the identification of plants depends on vein morphology. Altuntaş et al. [35] proposed a novel architecture for sorting maize seeds using CNN. Rahnemoonfar and Sheppard [36] suggest an automatic yield estimation system for the tomato plant based on robotic agriculture. Knoll et al. [37] propose self-learning CNN, which uses real-time visual sensor data to differentiate individual classes of plants.

11.2.4 Precision Livestock Farming

Precision livestock farming approaches are special topics as it is considered one of the issues of modern agriculture, as it is a part of precision farming. Several techniques are used in monitoring animal health indicators, which include comfort, pose estimation and behavior detection, and other factors of production. Gorczyca et al. [42], using a machine learning algorithm, predicted hair-coat temperature, piglet's skin, and core. Yukun et al. [44] propose a simple way to measure physical condition scores depending on deep learning and machine learning. Kvam [45] suggested an approach to calculate the Intra Muscular Fat (IMF) from ultrasound pictures. Zhang et al. [46] generate a real-time sow identification of behavior deep learning-based algorithm.

11.2.5 Pest Recognition

Although some insect species severely affect agricultural production and products, even if they are economically efficient. Recognition of an object is not just like a pest identification; it is a more complicated process that needs extra attention. Cheng et al. [47], in a complex background, used deep residual learning to recognize pests. In Partel et al. [49] artificial intelligence is used to create an automated vision-based technology to control pests such as the Asian citrus psyllid. Dawei et al. [51] suggested image-based transfer learning techniques for pest recognition. Li et al. [52] suggested a data augmentation method for the identification of pests and localization in the field by using CNN.

11.3 Background Details

11.3.1 Deep Learning

Deep learning comes under machine learning and works the same as ANN. Deep learning extracts relevant information from unstructured data. The main idea behind deep learning is that it uses the neural network layers for the learning features and analyzing the data. Further, the data features are extracted using different hidden layers present in the network. Every layer presented in the network is regarded as the perceptron, which is used for extracting low-level features. Hence, these high-level features are combined with low-level features, which reduces the local minimum problem. Therefore, with the recent advances in the field, deep learning has highly leveraged researchers.

DL can solve more complex problems fast because of its complex structure. However, executing the collections of datasets required to execute deep learning models, so many steps are involved.

This diagram shows the deep learning implementation by first taking the dataset's collection and dividing it into a test, train and validation set. In most cases, while doing the experimentation work, 80 percent of the training dataset and 20 percent of the test dataset is used. In recent times, with the help of transfer learning techniques, pre-trained models are used for doing the experimenting work. Moreover, to check the model significance, training and validation plots are used widely. A combination of popular performance metrics is used to classify images. Therefore, the evaluated studies are that to identify and classify the images, visualization techniques are applied.

Artificial intelligence was discovered to work when a list of mathematical and logical rules could describe a mentally challenging problem for a human. Artificial intelligence has rapidly expanded worldwide in recent years. Researchers have increasingly focused their efforts on applying experience, identifying the pattern in sound and images and making intuitive decisions. However, by capturing the information from a massive amount of information, ML algorithms improve prediction performance instead of manually creating rules based on the data analysis. The proposed approach allows for evidence-based decision-making that has made more effective. Machine learning uses supervised, unsupervised, reinforced and hybrid learning to extract meaningful relationships from data.

The principles of deep learning depend on artificial neural networks. However, in contrast to neural networks, deep neural networks have a deeper structure. Several machine learning approaches were used in shallow architecture before the big data era. Shallow architectures are useful for managing well-structured challenges, but they are not good enough for real-world applications like images, natural voice, human speech and language. Deep learning can manage these data.

Such input can still be processed by only a single layer ANN that has been modified as recent architecture. Deeper architecture is needed for complex data processing. Over the past few years, deep learning has shown its presence in the field of computer vision.

An artificial neural network can be created by interacting with a schematic representation of artificial neurons. During this activation function, the complete resultant is transmitted to the output. They can transfer the data obtained from one artificial unit's output layer of another artificial unit. As mathematical expressions, inputs are represented as X1, X2, X3 … Xn. The weights indicate how efficiently the incoming data is passed via the inputs to the output. W1, W2, W3 … Wn is the mathematical expression for weights. The sum function is used to calculate input by combining each intake value with weight.

However, the addition of each received input is further weight multiplication is performed by its weights. The additional function is defined mathematically as:

$$NetInput = Xwixi = w1x1 + w2x2 + \ldots + wixi \qquad (11.1)$$

The activation function controls the artificial nerve unit's output. The most common activation functions are softmax, sigmoid, and threshold. Data is transmitted to the output depending upon what activation function is selected. Depending on the activation function, a feedforward network generates a value sent to the other cell or the outside world. During the feedback network, the input value and output value are transmitted simultaneously.

A significant amount of data is essential for network performance during the process of training. However, a large volume of data allows various techniques for enhancing artificial neural network learning performances [8]. It seems to be more difficult to train the shallow system, according to research. Furthermore, as the architecture becomes much more complicated, the training period must be extended. Literature has provided some ideas about how to identify these issues [16]. Despite its simple structure, deep learning has a widely accepted rectifier linear unit that produces very valuable research observations. It returns zero for negative numbers and raises linearly for positive amounts [17]. The convergence of the estimates to the desired output is facilitated by this activation function. ReLU is a useful function in deep networks because of its shape, as it is reasonably cheap to compute and free from the problem of vanishing gradient.[15] Additionally, the ReLU activation function has several drawbacks, which were subsequently addressed by the development of Softplus, ELU, leaky ReLU, and PReLU swish activation function. Furthermore, a number of other algorithms were developed to improve these algorithms, including Nadam, Adadelta, Adam, and Adagrad, RMSprop, Adamax, AMSGrad, and Nesterov accelerated gradient algorithm [18]. As mentioned above, when deeper networks are being used, the challenges are only part of the work that has to be done. In deep networks, model selection has always been a challenge since the model must be chosen in a way that suits the problem sufficiently. However, an insufficient data structure or an overfit to the model heavily influences predictions. A tradeoff between the bias and the variance is essential to avoid overfitting and inadequacies regarding learning from the network [19]. To alleviate the problem of overfitting large volume of data can be gathered and used to modify the model. If it is impossible to collect current data, the data augmentation techniques are used to improve the existing training set. Training set also alleviates overfitting issues, in addition to improving the data. Furthermore, observing the validation set's performance can help us determine when the training should stop. It is also possible to reduce overfitting of the network by applying regulation and dropout.

11.3.2 Motivation of this Study

The basic steps in farming involve disease detection and selecting an appropriate solution for controlling that crop's diseases, which farmers have used for generations to prevent crop loss and maintain quality. Moreover, as a result of global climate change, modifying people's lifestyles, pollution and various other causes, many crop diseases have increased significantly. Farmers need to know about these plant diseases to determine and control them. Furthermore, it is difficult for a farmer to be aware of all the diseases due to the different diseases. Additionally, with the vast scale of farming, it is both economically and physically impossible for a farmer to monitor and control crops. Furthermore, a machine learning-based approach can be utilized for reliable and real-time identification and classification of diseases in crops, improving crop yield and quality while reducing labor cost and increasing farmer accuracy in crop cultivation.

11.3.3 Contribution

The following are the study's main contributions: Several infections and diseases have been discovered in various crops like tomatoes, potatoes, rice, apple and other fruits, vegetables as well as their symptoms are discussed in order to classify them. The procedures involved in automatic detection are discussed, as well as several strategies and algorithms. Furthermore, various types of possible diseases are present in several plants like tomato, potato, apple, rice and others are described as well as the symptoms that can use to identify and classify them. Image acquisition, image segmentation, image extraction, and disease classification are some of the generic phases in ML and DL that are used to classify and detect plant diseases automatically.

11.3.3.1 Similarity of Different Types of Plant Disease

Plant diseases are classified into two groups: biotic and abiotic. Diseases caused by living organisms like fungi, viruses and bacteria are called biotic diseases, whereas plants' abiotic diseases are caused by excessively high temperature, bad weather, excess precipitation, insufficient vitamins, greenhouse gases and poor soil pH.

11.3.3.2 Steps Involved in Plant Disease Detection

The labor cost associated with the close and constant observation of crops for probable infection will also be reduced by the automated method of recognizing plant diseases. However, using machine learning techniques, researchers can detect diseases in crops and plants. Computer vision entails a generalized series of steps that range from procuring and collecting the images of plants by using various IoT sensing devices spread in the farm field to processing the images that were taken in order to feed into a disease categorization machine learning model.

11.3.3.3 Deep Learning in Tomato Diseases

Tomatoes are one of the most popular fruits, particularly in India. Vitamin C, Vitamin E, and beta carotene are among their components. They are also high in potassium, an important mineral that promotes health. It is a famous fruit that is full of nutrients. Although, bacteria, germs, and pests are regularly infected with various diseases (Table 11.2).

11.3.3.4 Deep Learning in Potato Diseases

The production of potatoes is the fourth highest among crops and one of the most plagued crops in agriculture. Various research papers in the literature focusing on the diagnosis of potato crop disease are discussed in the Table 11.3.

11.3.3.5 Deep Learning in Apple Diseases

Apples are a mineral-rich fruit that is widely cultivated around the world. Currently, there are more than 3,000 different kinds of apple. Apples are a common economic crop

TABLE 11.2

Classification of Tomato Diseases

Author Name	Year	Disease	Dataset	Classifier	Ref.
AS Chakravarthy, S Raman	2020	Early blight	PlantVillage dataset generated 4,281 tomato leaves. There were 100 images utilized for testing, and 4,141 images were used for training.	ResNet, Xception and other CNN-based transfer learning models	[53]
M Govardhan, M Veena	2019	Late and early blight, Mite, and Target spot	PlantVillage Dataset produced 1120 images of tomato leaves	Random Forest, SVM	[54]
K Balakrishna, M Rao	2019	Healthy, spider mites, powdery mildew, Verticillium wilt, leaf miners, Septoria leaf spot	600 images of tomato leaves that I created myself	K-Nearest Neighbor (KNN) and Probabilistic Neural Network (PNN) are based on neural networks	[55]
CU Kumari, SJ Prasad, G Mounika	2019	Leaf mold, Septoria leaf spot	PlantVillage dataset	Neural network involving backpropagation	[56]
P Tm, A Pranathi, K SaiAshritha, NB Chittaragi, SG Koolagudi	2018	Tomato leaves are categorized into ten disease types	PlantVillage dataset includes 18,160 images	CNN	[84]
J Shijie, J Peiyi, H Siping	2017	Low temperature and nutritional deficiency, gray mold, leaf mold, powdery mildew, white fly, plague of leaf miner	After image augmentation, a total of 43,398 images samples were created, and 5000 images were collected around the Korean Peninsula	Faster RCNN (FRCNN), Random Forest and Convolutional Neural Network (RFCNN), (Single Shot Detector (SSD))	[58]
M Brahimi, K Boukhalfa, A Moussaoui	2017	Tomato leaves have been found to have nine diseases	PlantVillage dataset 14,828 images are collected of tomato leaves	SVM, Random Forest, AlexNet, GoogLeNet	[59]

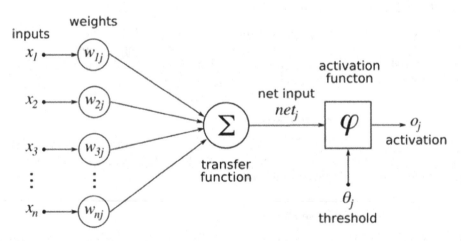

FIGURE 11.2
Diagrammatic representation of ANN.

TABLE 11.3

Classification of Potato Diseases

Author Name	Year	Disease	Dataset	Classifier	Ref.
U Suttapakti, A Bunpeng	2019	Early blight, late blight, and healthy potato leaves	PlantVillage dataset included 300 images of potato leaves.	Maximum-minimum color difference and Euclidian distance classification	[61]
S Arya, R Singh	2019	Early blight infected and healthy	PlantVillage dataset produced 4,004 images.	AlexNet model proposed CNN and transfer learning	[62]
M Al-Amin, TA Bushra, MN Hoq,	2019	Early blight, late blight and healthy	PlantVillage dataset collected 2,250 images of potato leaves.	Apply DCNN	[85]
M Islam, A Dinh, K Wahid, P Bhowmik	2017	Early blight, late blight, and healthy	PlantVillage dataset generated 300 potato leaves.	Using multi-class SVM	[63]
S Biswas, B Jagyasi, RP Singh, M Lal	2014	Late blight	27 images collected 340 regions of interest (ROI) after applying fuzzy c-means (FCM) (ROI)	Neural network using backpropagation	[64]

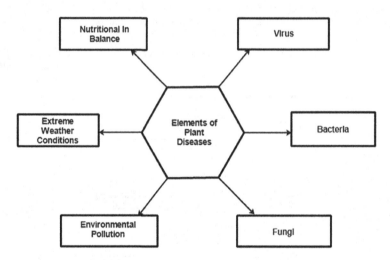

FIGURE 11.3
Causes of plant disease.

with significant nutritional and medical values. On the other hand, apples are frequently infected with various diseases, reducing yields and production (Table 11.4).

11.3.3.6 Deep Learning Approaches for High Spectral Images in Agricultural Field

Deep learning models have revolutionized this field with the recent advances in machine learning. So many deep learning techniques for problem solving, such as the classification of images in agriculture, have been suggested. Table 11.5 describes a few recent DL models for different data analytics in the smart agricultural field.

TABLE 11.4

Classification of Apple Diseases

Author Name	Year	Disease	Dataset	Classifier	Ref.
SV Militante, BD Gerardo, NV Dionisio	2019	Black rot	With 35,000 total images, the PlantVillage dataset was used.	Using convolutional neural network	[66]
P Jiang, Y Chen, B Liu, D He, C Liang	2019	Brownish spot, mosaic rust spot	To develop a disease dataset of apple leaf, 2,029 images of five different types of apple disease were collected. Image augmentation techniques were used to generate 26,377 images.	Transfer learning model based on DCNN, VGGNet, and SSD	[67]
Z Chuanlei, Z Shanwen, Y Jucheng, S Yancui, C Jia	2017	Mosaic and apple rust, powdery mildew	For each disease group, 90 images were collected.	Using SVM	[69]
BJ Samajpati, SD Degadwala	2016	Apple blotch, scab, and rot, as well as normal apples	There are 320 images in total, with 80 images from every class.	Using Random Forest algorithm	[70]
SR Dubey, AS Jalal	2016	Normal, apple scab, apple rot, and apple blotch apples	320 photos taken from Google images, with 80 images for every class.	Using multi-class SVM	[71]
S Dewliya, P Singh	2015	Rot in apple, scrub in apple	Not mentioned.	SVM with multiple classes and different kernels	[72]
A Awate, D Deshmankar, G Amrutkar, U Bagul, S Sonavane	2015	Apple blotch, apple scab, apple rot	Not mentioned.	Using ANN	[73]

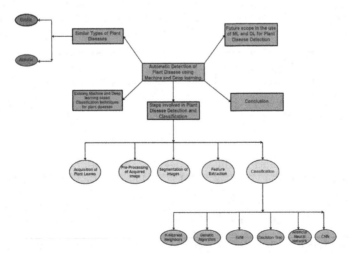

FIGURE 11.4
Automatic plant disease detection steps.

TABLE 11.5

Deep Learning Approach for High Spectral Images

Author Name	Agriculture Purpose	Techniques	Features	Ref.
L-C Chen et al.	Segmentation of semantic images	Deep convolutional networks used.	Semantic segmentation using deep learning.	[74]
A Ferreira et al.	Identification of drug crops	CNN classifiers in a pair.	A novel ML method for identifying drug crops has been developed.	[76]
C Deng et al.	Identification of hyperspectral image	CNN is used with virtual auto encoder.	The suggested strategy performed well as compared to cutting-edge techniques.	[77]
Z Niu et al.	Hyperspectral image classification	A spectral-spatial deep learning framework was used to identify hyperspectral images.	Deep learning techniques suited to small-scale classes.	[78]
Z Li, G Chen, T Zhang	Classification of crop	Architecture with a hybrid CNN and a transformer.	The CNN transformer exceeds the traditional approach by a large margin.	[79]
N Laban et al.	Coverages of land	Deep convolutional neural network monitoring land cover.	Crop and land cover classification accuracy was evaluated at 89% by DCNN.	[80]
Y Guo, X Jia, D Paull	Building a map of rice variety distribution	Deep convolutional neural networks for spectral and temporal domain.	With DCNN, accuracy was 98.87%, and with SVM, it was 57.49%.	[81]
Z Xin et al.	Cadmium residue in lettuce leaves can be predicted.	Least squares (partial) deep learning and SVM regression with layered auto encoders.	There's a lot of promise in a deep learning method for identifying heavy metal concentration in lettuce leaves.	[82]
Z Han, J Gao	Aflatoxin in peanuts has been identified	CNN architecture with five layers.	The recognition rate is 96% and 90%, respectively.	[83]

11.4 Conclusion

Plant diseases must be identified and classified correctly for crops to be grown successfully. Manual disease identification in crops faces many challenges since it requires significant finance, labor, and knowledge to correctly identify crop diseases. Furthermore, given elements like the size of the farm, several kinds of diseases and similar symptoms for various diseases harm crop yield and quality. So, based on the issues mentioned above, DL and ML are used to recognize plant diseases in various ways. Moreover, the algorithm for machine learning in the traditional sense, SVM, has shown extraordinary performance in classifying and identifying diseases in plants. In contrast, CNN has been demonstrated to be one of the most effective deep learning algorithms. According to the literature, much time and effort have already been spent forecasting the various plant diseases using their images. However, factors that must be highlighted include accuracy enhancement and real-time testing and deployment.

References

1. M. Ayaz, M. Ammad-Uddin, Z. Sharif, A. Mansour, and E.-H.-M. Aggoune, "Internet-of-Things (IoT)-based smart agriculture: Toward making the fields talk," *IEEE Access*, vol. 7, pp. 129551–129583, 2019.

2. Sciforce. Smart Farming: The Future of Agriculture. Accessed: Dec. 2, 2019. [Online]. Available: https://www.iotforall.com/smartfarming-future-of-agriculture/

3. Schuttelaar-partners.com. Smart Farming is Key for the Future of Agriculture. Accessed: Dec. 14, 2019. [Online]. Available: https://www.schuttelaar-partners.com/news/2017/smart-farming-is-key-for-thefuture-of-agriculture

4. R. Varghese and S. Sharma, "Affordable smart farming using IoT and machine learning," in Proc. 2nd Int. Conf. Intell. Comput. Control Syst. (ICICCS), Jun. 2018, pp. 645–650.

5. I. Goodfellow, Y. Bengio, and A. Courville, *Deep Learning*. Cambridge, MA: MIT Press, 2016, pp. 1–15.

6. S. Raschka and V. Mirjalili, *Machine Learning and Deep Learning With Python, Scikit-Learn and TensorFlow*. Birmingham: Packt Publishing, 2017, pp. 2–6.

7. C. Fyfe, "Artificial neural networks and information theory," PhD dissertation, Dept. Comput. Inf. Syst., University of Paisley, 2000.

8. L. Deng and D. Yu, "Deep learning: Methods and applications," *Found. Trends Signal Process.*, vol. 7, no. 3–4, pp. 197–387, 2014.

9. J. Heaton, "Artificial intelligence for humans," in *Neural Networks and Deep Learning*, vol. 3. Chesterfield, MO, USA: Heaton Research, 2015, pp. 165–180.

10. D. Graupe, *Principles of Artificial Neural Networks*. Chicago, IL: World Scientific, 2007, pp. 1–12.

11. B. Kröse and P. Smagt, *An Introduction to Neural Networks*. Amsterdam: Univ. of Amsterdam, 1996, pp. 15–20.

12. M. O. Ögücü, "Yapay sinir aˇglari ile sistem tanıma," M.S. thesis, Inst. Sci. Technol., Istanbul Teknik Univ., Istanbul, 2006.

13. J. A. Freeman and D. M. Skapura, *Neural Networks Algorithms, Applications, and Programming Techniques*. Reading, MA: Addison Wesley, 1991, pp. 18–50.

14. Ş. Emir, "Classification performance comparison of artificial neural networks and support vector machines methods: An empirical study on predicting stock market index movement direction," PhD dissertation, Inst. Social Sci., Istanbul Univ., Istanbul, 2013.

15. A. Sharma. Understanding Activation Functions in Deep Learning. Accessed: Dec. 1, 2019. [Online]. Available: https://www.learnopencv.com/understanding-activation-functions-in-deep-learning/

16. Y. Bengio, *Learning Deep Architectures for AI. Foundations and Trends in Machine Learning*, vol. 2, no. 1. Boston, MA: Nov. 2009, pp. 1–127.

17. A. Gulli and S. Pal, *Deep Learning With Keras*. Birmingham: Packt Publishing, 2017, pp. 68–90.

18. M. Toussaint. Introduction to Optimization. Second Order Optimization Methods. Accessed: Oct. 1, 2019. [Online]. Available: https://ipvs.informatik.uni-stuttgart.de/mlr/marc/teaching/13-Optimization/04-secondOrderOpt.pdf

19. J. Patterson and A. Gibson, *Deep Learning: A Practitioner's Approach*. Newton, MA: O'Reilly Media, 2017, pp. 102–121.

20. M. Kerkech, A. Hafiane, and R. Canals, "Deep learning approach with colorimetric spaces and vegetation indices for vine diseases detection in UAV images," *Comput. Electron. Agricult.*, vol. 155, pp. 237–243, Dec. 2018.

21. J. Lu, J. Hu, G. Zhao, F. Mei, and C. Zhang, "An in-field automatic wheat disease diagnosis system," *Comput. Electron. Agricult.*, vol. 142, pp. 369–379, Nov. 2017.

22. G. Hu, H. Wu, Y. Zhang, and M. Wan, "A low shot learning method for tea leaf's disease identification," *Comput. Electron. Agricult.*, vol. 163, Art. no. 104852, Aug. 2019.

23. S. Coulibaly, B. Kamsu-Foguem, D. Kamissoko, and D. Traore, "Deep neural networks with transfer learning in millet crop images," *Comput. Ind.*, vol. 108, pp. 115–120, Jun. 2019.

24. A. Cruz, Y. Ampatzidis, R. Pierro, A. Materazzi, A. Panattoni, L. De Bellis, and A. Luvisi, "Detection of grapevine yellows symptoms in vitis vinifera L. With artificial intelligence," *Comput. Electron. Agricult.*, vol. 157, pp. 63–76, Feb. 2019.

25. A. Picon, A. Alvarez-Gila, M. Seitz, A. Ortiz-Barredo, J. Echazarra, and A. Johannes, "Deep convolutional neural networks for mobile capture device-based crop disease classification in the wild," *Comput. Electron. Agricult.*, vol. 161, pp. 280–290, Jun. 2019.

26. N. Kussul, M. Lavreniuk, A. Kolotii, S. Skakun, O. Rakoid, and L. Shumilo, "A workflow for sustainable development goals indicators assessment based on high-resolution satellite data," *Int. J. Digit. Earth*, vol. 13, no. 2, pp. 309–321, 2019.

27. V. Persello, A. Tolpekin, J. R. Bergado, and R. A. de By, "Delineation of agricultural fields in smallholder farms from satellite images using fully convolutional networks and combinatorial grouping," *Remote Sens. Environ.*, vol. 231, Art. no. 111253, Sep. 2019.

28. Y. Zhou, J. Luo, L. Feng, Y. Yang, Y. Chen, and W. Wu, "Longshort-term-memory-based crop classification using high-resolution optical images and multi-temporal SAR data," *GISci. Remote Sens.*, vol. 56, no. 8, pp. 1170–1191, Nov. 2019.

29. Q. Yang, L. Shi, J. Han, Y. Zha, and P. Zhu, "Deep convolutional neural networks for Rice grain yield estimation at the ripening stage using UAVbased remotely sensed images," *Field Crops Res.*, vol. 235, pp. 142–153, Apr. 2019.

30. S. Zhao, X. Liu, C. Ding, S. Liu, C. Wu, and L. Wu, "Mapping rice paddies in complex land-scapes with convolutional neural networks and phenological metrics," *GISci. Remote Sens.*, vol. 57, no. 1, pp. 37–48, 2019.

31. J. Dyson, A. Mancini, E. Frontoni, and P. Zingaretti, "Deep learning for soil and crop segmenta-tion from remotely sensed data," *Remote Sens.*, vol. 11, no. 16, p. 1859, Aug. 2019.

32. P. Nevavuori, N. Narra, and T. Lipping, "Crop yield prediction with deep convolutional neural networks," *Comput. Electron. Agricult.*, vol. 163, Art. no. 104859, Aug. 2019.

33. G. L. Grinblat, L. C. Uzal, M. G. Larese, and P. M. Granitto, "Deep learning for plant identification using vein morphological patterns," *Comput. Electron. Agricult.*, vol. 127, pp. 418–424, Sep. 2016.

34. B. Veeramani, J. W. Raymond, and P. Chanda, "DeepSort: Deep convolutional networks for sorting haploid maize seeds," *BMC Bioinf.*, vol. 19, no. S9, pp. 85–93, Aug. 2018.

35. Y. Altuntaş, Z. Cömert, and A. F. Kocamaz, "Identification of haploid and diploid maize seeds using convolutional neural networks and a transfer learning approach," *Comput. Electron. Agricult.*, vol. 163, Art. no. 104874, Aug. 2019.

36. M. Rahnemoonfar and C. Sheppard, "Deep count: Fruit counting based on deep simulated learning," *Sensors*, vol. 17, no. 4, p. 905, Apr. 2017.

37. F. J. Knoll, V. Czymmek, S. Poczihoski, T. Holtorf, and S. Hussmann, "Improving efficiency of organic farming by using a deep learning classification approach," *Comput. Electron. Agricult.*, vol. 153, pp. 347–356, Oct. 2018.

38. N. Häni, P. Roy, and V. Isler, "A comparative study of fruit detection and counting methods for yield mapping in apple orchards," *J. Field Robot.*, vol. 37, pp. 1–20, Aug. 2019.

39. Y. Tian, G. Yang, Z. Wang, H. Wang, E. Li, and Z. Liang, "Apple detection during different growth stages in orchards using the improved YOLO-V3 model," *Compute. Electron. Agricult.*, vol. 157, pp. 417–426, Feb. 2019.

40. J. Gené-Mola, V. Vilaplana, J. R. Rosell-Polo, J.-R. Morros, J. Ruiz-Hidalgo, and E. Gregorio, "Multi-modal deep learning for fuji apple detection using RGB-D cameras and their radiomet-ric capabilities," *Comput. Electron. Agricult.*, vol. 162, pp. 689–698, Jul. 2019.

41. H. Kang and C. Chen, "Fruit detection and segmentation for apple harvesting using visual sen-sor in orchards," *Sensors*, vol. 19, no. 20, p. 4599, Oct. 2019.

42. M. T. Gorczyca, H. F. M. Milan, A. S. C. Maia, and K. G. Gebremedhin, "Machine learning algo-rithms to predict core, skin, and hair-coat temperatures of piglets," *Comput. Electron. Agricult.*, vol. 151, pp. 286–294, Aug. 2018.

43. X. Huang, Z. Hu, X. Wang, X. Yang, J. Zhang, and D. Shi, "An improved single shot multibox detector method applied in body condition score for dairy cows," *Animals*, vol. 9, no. 7, p. 470, Jul. 2019.

44. S. Yukun, H. Pengju, W. Yujie, C. Ziqi, L. Yang, D. Baisheng, L. Runze, and Z. Yonggen, "Automatic monitoring system for individual dairy cows based on a deep learning framework that provides identification via body parts and estimation of body condition score," *J. Dairy Sci.*, vol. 102, no. 11, pp. 10140–10151, Nov. 2019.

45. J. Kvam and J. Kongsro, "In vivo prediction of intramuscular fat using ultrasound and deep learning," *Comput. Electron. Agricult.*, vol. 142, pp. 521–523, Nov. 2017.

46. Y. Zhang, J. Cai, D. Xiao, Z. Li, and B. Xiong, "Real-time sow behavior detection based on deep learning," *Comput. Electron. Agricult.*, vol. 163, Art. no. 104884, Aug. 2019.

47. X. Cheng, Y. Zhang, Y. Chen, Y. Wu, and Y. Yue, "Pest identification via deep residual learning in complex background," *Comput. Electron. Agricult.*, vol. 141, pp. 351–356, Sep. 2017.

48. Y. Shen, H. Zhou, J. Li, F. Jian, and D. S. Jayas, "Detection of storedgrain insects using deep learning," *Comput. Electron. Agricult.*, vol. 145, pp. 319–325, Feb. 2018.

49. V. Partel, L. Nunes, P. Stansly, and Y. Ampatzidis, "Automated visionbased system for monitoring Asian citrus psyllid in orchards utilizing artificial intelligence," *Comput. Electron. Agricult.*, vol. 162, pp. 328–336, Jul. 2019.

50. K. Thenmozhi and U. Srinivasulu Reddy, "Crop pest classification based on deep convolutional neural network and transfer learning," *Comput. Electron. Agricult.*, vol. 164, Art. no. 104906, Sep. 2019.

51. W. Dawei, D. Limiao, N. Jiangong, G. Jiyue, Z. Hongfei, and H. Zhongzhi, "Recognition pest by image-based transfer learning," *J. Sci. Food Agricult.*, vol. 99., no. 10, pp. 4524–4531, 2019.

52. Li, X. Jia, M. Hu, M. Zhou, D. Li, W. Liu, R. Wang, J. Zhang, C. Xie, L. Liu, F. Wang, H. Chen, T. Chen, and H. Hu, "An effective data augmentation strategy for CNN-based pest localization and recognition in the field," *IEEE Access*, vol. 7, pp. 160274–160283, 2019.

53. A. S. Chakravarthy and S. Raman (2020) "Early blight identification in tomato leaves using deep learning," In *2020 International Conference on Contemporary Computing and Applications (IC3A)*, IEEE, 154–158.

54. M. Govardhan and M. Veena (2019) "Diagnosis of tomato plant diseases using random forest," In *2019 Global Conference for Advancement in Technology (GCAT)*, IEEE, 1–5.

55. K. Balakrishna and M. Rao "Tomato plant leaves disease classification using KNN and PNN," *Int J Computer Vision Image Process*, vol. 9, no. 1, pp. 51–63, 2019.

56. C. U. Kumari, S. J. Prasad and G. Mounika (2019) "Leaf disease detection: feature extraction with K-means clustering and classification with ANN," In *2019 3rd International conference on computing methodologies and communication (ICCMC)*, IEEE, 1095–1098.

57. W. Jin, Z. J. Li, L. S. Wei, and H. Zhen (2000) "The improvements of BP neural network learning algorithm," In *WCC 2000-ICSP 2000. 2000 5th international conference on signal processing proceedings. 16th world computer congress*, IEEE, vol. 3, 1647–1649.

58. J. Shijie, J. Peiyi, and H. Siping (2017) "Automatic detection of tomato diseases and pests based on leaf images," In *2017 Chinese automation congress (CAC)*, IEEE, 2537–2610.

59. M. Brahimi, K. Boukhalfa, and A. Moussaoui, "Deep learning for tomato diseases: classification and symptoms visualization," *Appl. Artif. Intell.* vol. 31 no.4, pp. 299–315, 2017.

60. L. Liu and G. Zhou (2009) "Extraction of the rice leaf disease image based on BP neural network," In *2009 International Conference on Computational Intelligence and Software Engineering*, IEEE, 1–3.

61. U. Suttapakti and A. Bunpeng (2019) "Potato leaf disease classification based on distinct color and texture feature extraction," In *2019 19th International Symposium on Communications and Information Technologies (ISCIT)*, IEEE, 82–85.

62. S. Arya and R. Singh (2019) "A comparative study of CNN and AlexNet for detection of disease in potato and mango leaf," In *2019 International Conference on Issues and Challenges in Intelligent Computing Techniques (ICICT)*, IEEE, vol. 1, 1–6.

63. M. Islam, A. Dinh, K. Wahid, and P. Bhowmik (2017) "Detection of potato diseases using image segmentation and multiclass support vector machine," In *2017 IEEE 30th Canadian conference on electrical and computer engineering (CCECE)*, IEEE, 1–4.

64. S. Biswas, B. Jagyasi, B. P. Singh, and M. Lal "Severity identification of potato late blight disease from crop images captured under uncontrolled environment," In *2014 IEEE Canada international humanitarian technology conference-(IHTC)*, IEEE, 1–5.

65. M. J. Hasan, S. Mahbub, M. S. Alom, and M. A. Nasim (2019) "Rice disease identification and classification by integrating support vector machine with deep convolutional neural network," In *2019 1st International Conference on Advances in Science, Engineering and Robotics Technology (ICASERT)*, IEEE, 1–6.

66. S. V. Militante, B. D. Gerardo, and N. V. Dionisio (2019) "Plant leaf detection and disease recognition using deep learning," In *2019 IEEE Eurasia conference on IOT, communication and engineering (ECICE)*, IEEE, 579–582.

67. P. Jiang, Y. Chen, B. Liu, D. He, and C. Liang, "Real-time detection of apple leaf diseases using deep learning approach based on improved convolutional neural networks," *IEEE Access*, vol. 7, pp. 59069–59080, 2019.

68. C. Szegedy et al.,(2015) "Going deeper with convolutions," In *Proceedings of the IEEE conference on computer vision and pattern recognition*, 1–9.

69. Z. Chuanlei, Z. Shanwen, Y. Jucheng, S. Yancui, and C. Jia, "Apple leaf disease identification using genetic algorithm and correlation based feature selection method," *Int J Agricul Biolog Eng*, vol. 10, no. 2, pp. 74–83, 2017.

70. B. J. Samajpati and S. D. Degadwala (2016) "Hybrid approach for apple fruit diseases detection and classification using random forest classifier," In *2016 International conference on communication and signal processing (ICCSP)*, IEEE, 1015–1019.

71. S. R. Dubey and A. S. Jalal, "Apple disease classification using color, texture and shape features from images," *SIViP*, vol. 10, no. 5, pp. 819–826, 2016.

72. S. Dewliya and P. Singh, "Detection and classification for apple fruit diseases using support vector machine and chain code," *Inter Res J Eng Technol (IRJET)*, vol. 2, no. 04, pp. 2097–2104, 2015.

73. A. Awate, D. Deshmankar, G. Amrutkar, U. Bagul, and S. Sonavane (2015) "Fruit disease detection using color, texture analysis and ANN," In *2015 International Conference on Green Computing and Internet of Things (ICGCIoT)*, IEEE, 970–975.

74. L.-C. Chen, G. Papandreou, I. Kokkinos, K. Murphy, and A. L. Yuille, "Semantic image segmentation with deep convolutional nets and fully connected CRFs," 2014, arXiv:1412.7062. [Online]. Available: http://arxiv.org/abs/1412.7062

75. Y. Guo, H. Cao, J. Bai, and Y. Bai, "High efficient deep feature extraction and classification of spectral-spatial hyperspectral image using cross domain convolutional neural networks," *IEEE J. Sel. Topics Appl. Earth Observ. Remote Sens.*, vol. 12, no. 1, pp. 345–356, Jan. 2019.

76. A. Ferreira, S. C. Felipussi, R. Pires, S. Avila, G. Santos, J. Lambert, J. Huang, and A. Rocha, "Eyes in the skies: A data-driven fusion approach to identifying drug crops from remote sensing images," *IEEE J. Sel. Topics Appl. Earth Observ.Remote Sens.*, vol. 12, no. 12, pp. 4773–4786, Dec. 2019.

77. C. Deng, Y. Xue, X. Liu, C. Li, and D. Tao, "Active transfer learning network: A unified deep joint spectral–spatial feature learning model for hyperspectral image classification," *IEEE Trans. Geosci. Remote Sens.*, vol. 57, no. 3, pp. 1741–1754, Mar. 2019.

78. Z. Niu, W. Liu, J. Zhao, and G. Jiang, "DeepLab-based spatial feature extraction for hyperspectral image classification," *IEEE Geosci. Remote Sens. Lett.*, vol. 16, no. 2, pp. 251–255, Feb. 2019.

79. Z. Li, G. Chen, and T. Zhang, "A CNN-transformer hybrid approach for crop classification using multitemporal multisensor images," *IEEE J. Sel. Topics Appl. Earth Observ. Remote Sens.*, vol. 13, no. 13, pp. 847–858, 2020.

80. N. Laban, B. Abdellatif, H. M. Ebeid, H. A. Shedeed, and M. F. Tolba (Dec. 2018) "Seasonal multi-temporal pixel based crop types and land cover classification for satellite images using convolutional neural networks," in *Proc. 13th Int. Conf. Comput. Eng. Syst. (ICCES)*, 21–26.

81. Y. Guo, X. Jia, and D. Paull (Dec. 2018) "Mapping of rice varieties with Sentinel-2 data via deep CNN learning in spectral and time domains," in *Proc. Digit. Image Comput., Techn. Appl. (DICTA)*, 1–7.

82. Z. Xin, S. Jun, T. Yan, C. Quansheng, W. Xiaohong, and H. Yingying, "A deep learning based regression method on hyperspectral data for rapid prediction of cadmium residue in lettuce leaves," *Chemometric Intell. Lab. Syst.*, vol. 200, Art. no. 103996, May 2020.

83. Z. Han and J. Gao, "Pixel-level aflatoxin detecting based on deep learning and hyperspectral imaging," *Comput. Electron. Agricult.*, vol. 164, Art. no. 104888, Sep. 2019.

84. P. Tm, A. Pranathi, K. SaiAshritha, N. B. Chittaragi and S. G. Koolagudi, "Tomato Leaf Disease Detection Using Convolutional Neural Networks," *2018 Eleventh International Conference on Contemporary Computing (IC3)*, 2018, pp. 1–5. doi: 10.1109/IC3.2018.8530532.

85. M. Al-Amin, T. A. Bushra and M. Nazmul Hoq, "Prediction of Potato Disease from Leaves using Deep Convolution Neural Network towards a Digital Agricultural System," *2019 1st International Conference on Advances in Science, Engineering and Robotics Technology (ICASERT)*, 2019, pp. 1–5, doi: 10.1109/ICASERT.2019.8934933.

12

A Complete Framework for LULC Classification of Madurai Remote Sensing Images with Deep Learning-based Fusion Technique

T. Gladima Nisia

AAA College of Engineering and Technology, Sivakasi, India

S. Rajesh

Mepco Schlenk Engineering College, Sivakasi, India

CONTENTS

DOI: 10.1201/9781003206736-12

12.1 Introduction

Remote sensing image classification has been a trending research topic for many years, yet many improvements have been made by different researchers throughout these years. There are many different algorithms and different methodologies applied for classification. Some of the methods try to improve their performance by extracting features, some concentrate on feature selection or reduction, and some on the classification methodologies. But, most of the research forgets to concentrate on the image fusion area. Image fusion tries to integrate important necessary details from multiple sources of images and make it a single fused image. Images such as MS, PAN, and hyperspectral (HS) are used for fusion [1]. The main aim of the fusion is to create an increased resolution image in spatial and spectral aspects. The benefit of multispectral and panchromatic images are high spectral and spatial resolution, respectively [2]. Numerous approaches have been proposed for the fusion of remote sensing images [3].

Different approaches are handled to bring out the best fusion image. The proposed method in [4] works using the adaptive IHS method. Then repeated filtering is carried out using a multiscale guided filter. Here, the process is carried out by obtaining the MS image's Intensity component. Then extract detailed information from the PAN image, which is followed by information gain generation and the fused image is obtained finally. Another approach is based on the sparse model, which produces rich feature representation and better results [5]. Here, the analysis operator is trained directly using the input image and the geometric analysis operator learning method is used to up-sample multispectral and panchromatic images with low resolution. Unlike this method, Kai Zhang et al. used a sparse coding-based scheme of convolution structure sparse coding (CSSC) for image fusion [6]. The system merges convolution sparse coding and MS and PAN image relationships. Then, CSSC is elaborated by presenting structural sparsity. Finally, feature maps are calculated by substitute optimization, and the fused image is reconstructed.

The dictionary-based learning method is also used for the fusion of images [7]. The deep neural networks are involved in abstracting efficient features. Here, the MS and PAN images are combined with the help of CNN. The system handles the entire process with the help of the superpixel instead of a pixel. The CNN is introduced to extract joint feature representation based on local region representation using superpixel [8]. Based on the feature fusion, a complete learning framework was proposed, namely deep multiple instance learning, using the joint spatial-spectral details of panchromatic and multispectral images, respectively. The information of both of these is fused to come up with a fusion feature set. The feature set is fed into a network to understand the high-level feature fusion [9].

Compared to all existing works, our proposed system provides a very different and efficient way of handling image fusion and classification. Here, rather than using the input remote sensing image directly for feature extraction and classification, we will use the fused result of the MS image and PAN image. This approach results in improved classification accuracies. The highlights of the system proposed are four-fold, as listed below:

- The proposed system works with fused images. By doing so, the high spatial and high spectral resolutions problem is solved. Thus, the proposed system provides a greater start with an excellent input which is an image with greater quality.
- The proposed system utilizes the deep multi-features extracted from the CNN and combines those features with some other features to produce a multiple feature set. It came up with much better classification accuracy.
- Mainly, the existing works are classified using a single classifier, but the proposed system tries to build a combined classifier system (CCS) which could result in a combination of multiple member classifiers. Thus, resulting in greater accuracy.
- The experimental results and discussions are provided for the proposed approach, based on which it is clear that the proposed work yields much improved results when related to existing work.

The rest of this chapter is as follows: Section 12.2 delivers an outline of works related to image fusion, extraction of features and image classification. Section 12.3 provides the problem statement. Section 12.4 offers a detailed explanation of our proposed work. The experimental results and their evaluations are explained in Section 12.5. At the end, Section 12.6 gives the conclusion of the proposed work.

12.2 Related Work

The section deals with the research history related to the Linear Imaging Self-Scanning Sensors (LISS) IV images created. The LISS IV images are obtained from IRS-P6 satellite. These remote sensing images are used in many applications like urban planning [10], agriculture [11,12], preventing natural calamities [13] and so on. A detailed explanation of the IRS-P6 Landsat sensors is presented in [14]. Different methods are presented at various time periods for monitoring the land areas also [15]. Several works were done earlier, but the discussion will highlight some of the major contributions. The method proposed in [16] did enhancement for the image using a different filtering method and produces an improved enhanced image as its output.

12.2.1 Image Fusion

There are some proposed methods for image fusion applied to remote sensing images. An earlier approach [17] proposed a work that could fuse the images of LISS III and LISS IV to bring out an efficient fused image. The high spatial–low spectral resolution LISS IV image with high spectral–low spatial resolution LISS III image is fused to bring out a fused high spatial image with high spectral resolution. Then the classified output is attained by applying unsupervised techniques. Another work proposed gives the different resampling techniques for LISS III, LISS IV and Cartography + Satellite (CARTOSAT)-1. The work presents ways of preserving image quality by using different interpolation functions and resampling techniques [18].

12.2.2 Process of Feature Extraction

This particular sub-section discusses the works that are done related to feature extraction. Rajesh et al. [19] proposed a system that could extract deep features using CNN. A different network approach is employed here by using pre-defined filter values. Then the object-oriented features are added to the deep features. Thus, a new feature set is obtained, and the classification process is done with the new feature set. Rajesh et al. [20] proposed a system that extracts features like wavelet packet statistical and another set of features like wavelet packet co-occurrence. They are used for further classification. Arivazhagan and Ganesan [21] proposed a work that explains texture classification using different wavelet features: statistical, co-occurrence and combined statistical-co-occurrence features. The proposed method proved that the wavelet-based combined statistical-wavelet co-occurrence features produced a good classification accuracy.

12.2.3 Feature Selection

Sometimes trying to extract features from many methods lead to many numbers of features. Also, the processing of those features multiplies the complexity of the system. Thus, the features have to be reduced. Rather than reducing the features, the important features are selected for classification purposes. Rajesh et al. [22] made selections of features using a method that utilized the genetic algorithm. The system outperformed many existing methods like predictive data analysis (PDA) and linear discriminant analysis (LDA).

12.2.4 Classification of Images

The section explains the existing classification methods for LISS IV images. Rajesh [23] proposed a system that classifies the given image using the important features selected using a genetic algorithm. The system uses a hybrid learning method called adaptive network-based fuzzy inference system (ANFIS). The method uses artificial neural networks (ANN) and fuzzy logic. Rajesh et al. [19] proposed a system in which the object-oriented deep features are classified using CNN. The objects in the remote sensing image, such as land cover/land mapping, are identified and classified. Another work classifies the remote sensing (RS) image using the decision tree method [24]. Yet another method proposes a different way of classification with SVM (support vector machine) using multiple kernel function. The method tries to extract buildings from the complex urban scene using high-resolution satellite imagery [25].

12.3 Problem Statement

Remote sensing image classification has become one of the unavoidable research areas because of its useful applications in everyday life. Every pixel of the RS image contains thousands of details within itself. So, the existing systems try to classify the RS images using different classification methods. It also employs different ways in which the features of the image can be extracted. Rajesh et al. [19] show one of the ways of classifying the LISS IV image using the object-based approach. The Input LISS IV image is segmented into objects using a multi-resolution algorithm. The system then works by extracting deep features from the segmented image using a CNN. As another set, the texture features are also extracted. The extracted deep features are then combined with the texture features. Thus, as a result, the final feature set is obtained, which is used for classification and is done by CNN. The system presents an efficient classification method and results in improved classification accuracy. But, most of the research here does not concentrate on the image fusion technique. One of the best ways of improving the feature extraction and classification techniques is to use the fused image as the input image.

12.4 Proposed Work

12.4.1 System Overview

The proposed system is split into the following segments: image fusion, feature extraction, feature selection, and classification. The proposed system's general plan is to use MS and PAN images to generate a fused image, which is then used as input for the classification process. The architectural view of the proposed system is presented in Figure 12.1.

12.4.2 Image Fusion

The remote sensing fusion method consists of two main steps: extraction and fusion of features [26]. The proposed method uses convolutional kernels for feature extraction and to combine the MS and PAN images.

Network design:

Our proposed method introduces two different ways of feature extraction from the MS LISS IV and PAN LISS IV images individually. The network is fed with those two inputs. Finally, an output MS image with PAN image spatial resolution is obtained. While processing two images with the same CNN, the two images should be of the same dimensionality. Usually, the MS and PAN images obtained are not of the same dimensionality. So, the panchromatic image is brought to the size of the multispectral image by downsampling it. The fusion method consists of more layers in the CNN. The deeper the network becomes, the more the fusion

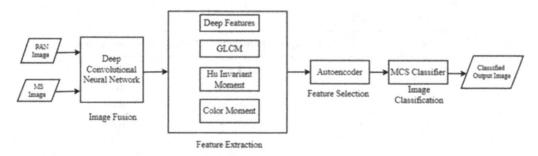

FIGURE 12.1
Architectural view of the proposed system.

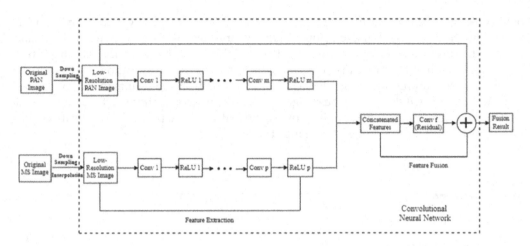

FIGURE 12.2
Proposed system for fusion of MS and PAN images.

results are improved. The deeper the network, high nonlinearities are exploited, giving out high-level features which are helpful for the fusion task. The network for the system proposed for fusion is given in Figure 12.2. The configuration of the CNN is mentioned in Table 12.1.

Throughout the process, the convolutional kernels are applied to extract features and, in the main thread, the convolutional kernels are used to fuse the extracted features. The automatic way of learning the kernels in CNN is enabled through deep learning techniques while training the network to generate improved image fusion output. The feature extraction is done through convolution and ReLU layer. The information about the low-resolution and high-resolution MS images is included in the mask. The mask is calculated in the last layer, and is referred to as residual learning. The mask is used for obtaining the final fused image by overlaying it over the low-resolution input image.

Network training:

The first step in training the CNN network is to select the optimal parameters. Here, the proposed system uses a different loss function to calculate the error and

TABLE 12.1

Configuration of CNN

Procedure	Branch	Layer	Kernel Size	Feature Map	Activation
Feature extraction	MS branch	1	$3 \times 3 \times 4$	64	ReLU
		m	$3 \times 3 \times 64$	32	ReLU
		others	$3 \times 3 \times 64$	64	ReLU
	PAN branch	1	$3 \times 3 \times 1$	64	ReLU
		P	$3 \times 3 \times 64$	32	ReLU
		others	$3 \times 3 \times 64$	64	ReLU
Fusion of features	Main thread	f	$3 \times 3 \times 64$	4	none

improve the training process. Let x1 and x2 represent the MS and PAN downsampled images. The input MS image is denoted as "y." The objective of the training is to derive an objective function, F: Y = f(x1, x2). The predicted MS image with high resolution is represented as Y. The residual image is created initially by finding the difference between the MS image's high and low resolution using Eq. (12.1)

$$\text{Residual Image}, r = Y - x1 \tag{12.1}$$

The residual image matrix consists of mostly zero or smaller value, resulting in a sparse residual image. Thus, the redundant information is ignored, and only the necessary features can be considered for improvement of the MS image's spatial resolution. Add low-resolution MS image with residual image, to bring out high-resolution MS image. The loss function is written as in Eq. (12.2).

$$L = \frac{1}{n} \sum_{i=1}^{n} || r^{(i)} - g(x_1^{(i)}, x_2^{(i)}) ||^2 \tag{12.2}$$

Here, $r^{(i)}$ represents the residual image and $g\left(x_1^{(i)}, x_2^{(i)}\right)$ refers to the obtained residual image. The residual learning is implemented by doing the following modifications to the loss layer of the network. The following are the parts of the loss layer: (1) MS image, (2) Residual image, and (3) MS high-resolution images. Now, add (1) and (2) to generate the predicted fusion result. The network is trained using the Eq. (12.3) with gradient descent algorithm and backpropagation concepts.

$$
\begin{aligned}
L &= \frac{1}{n} \sum_{i=1}^{n} || r^{(i)} - g\left(x_1^{(i)}, x_2^{(i)}\right)^2 || \\
&= \frac{1}{n} \sum_{i=1}^{n} || y^{(i)} - x_1^{(i)} - g\left(x_1^{(i)}, x_2^{(i)}\right)^2 || \\
&= \frac{1}{n} \sum_{i=1}^{n} || y^{(i)} - (x_1^{(i)} + g\left(x_1^{(i)}, x_2^{(i)}\right)^2 || \\
&= \frac{1}{n} \sum_{i=1}^{n} || y^{(i)} - f\left(x_1^{(i)}, x_2^{(i)}\right)^2 ||
\end{aligned}
\tag{12.3}
$$

The mini-batch with gradient descent algorithm with the method of backpropagation is used to train the network with the loss function given here in Eq. (12.3)

12.4.3 Feature Extraction

This is the process of expressing information in an image with low-dimension features. Extraction of a single feature may not be enough for efficient classification. So, when multiple features are extracted, the important information in the image is obtained without any loss. Multiple features to be extracted are deep features, shape feature, color feature and texture feature. The obtained feature sets are joined to obtain a single feature set.

12.4.3.1 Deep Features

To obtain the deep features, the system utilizes the available deep neural network, namely convolutional neural network (CNN). The input fused image is initially convolved with a filter matrix with randomly initialized values. The output of the convolution step is then fed into the pooling layer, which pools the convolution matrix values. The results are then processed using the ReLU layer. Here we present three layers of CNN, namely convolutional, pooling, and ReLU layers. The convolutional layer executes multiple times of convolution. The pooling layer downsamples the maximum of each block considered from the image. Stack the three layers to bring up a complete CNN architecture. The detailed working of the convolution and ReLU layers is explained below.

12.4.3.1.1 Convolutional Layer

1. Network:

Every pixel represents the detail of an image. If the fully connected network is employed, it may result in several parameters. Consider an RGB image. This image has parameters for each neuron. Thus, when the neural network is used, it may result in millions of parameters. A large number of parameters results in slowing down the whole process and may lead to overfitting. Based on the examination of images, it is understood that the features are local, and the low-level features are obtained easily. So, it is possible to decrease the fully connected network to a locally connected network. This is the basic idea of CNN.

Just like every other image processing does, a single block of an image is locally connected to a neuron. The block here represents the feature window. So, the parameter count can be reduced to a minimal amount, but the performance is not reduced. More features are extracted using the single blocks connected to another neuron. The layer's depth is based on how many times an area is connected to other neurons. For instance, if the same area is connected to five other neurons. Thus, the depth is five in the new layer.

All deep information of the network (for example, three channels of RGB) is connected to the following neuron to connect local information in height and width. When calculated based on this concept, there might be 5×5×5 parameters for the neuron if 5×5 window is used. In 5×4×3, 5 and 4 represent the height and width of the window, respectively, and 3 represents the depth of the layer.

The chosen window slides throughout the image matrix to perform convolution operation. Also, the height and width be a 2D one. For example, if the window slides one pixel per time, or stride one, in an image of size 32×32×3 and 5×5 window size, then there are 28×28×depth neurons. It is observed that the size of the image is decreasing by 4. In order to avoid this, zero padding is done at the edges. For example, if it is padded with two pixels, it will result in the 32×32×depth neurons, thus maintaining the image's size. Let's see the stride 1 case. If the window size is 'w,' then zero-pad with (w-1)/2 pixels. Thus, it is concluded that zero padding performs better than without zero padding.

The next important point here is "stride." The stride represents the shifting length of the window. For example, if the value of stride is 1 and window size is 3×3 in 7×7×3 image without zero padding, then there are 5×5× depth_size neurons. If the stride value is changed from 1 to 2 and other values remain the same, then there are 3×3× depth neurons in the next layer. Thus, it is concluded that there are $\left[\dfrac{W-w}{s}+1\right] X \left[\dfrac{H-w}{s}+1\right] X \, depth$ neurons in the next layer. Where 's' represents stride, 'w x w' represents window size in W × H image.

2. Activation function

In the earlier neuron networks, the sigmoid function is used as the activation function, whereas many other options are available. One of them is rectified linear units (ReLUs). The ReLU function is $f(x) = \max(0, x)$. This means that the value of a pixel is taken as such if the value is above '0' or else it is considered as '0.' When the ReLU function is compared with the performances of the sigmoid function, it is clearly understood that the network with ReLU layer took lesser iteration time with equal training errors. Thus most of the CNN models are implemented using the ReLU activation function.

12.4.3.1.2 *Pooling Layer*

There are still some other parameters used in the neural networks in addition to the locally connected network and parameter sharing. There may be chances of getting overfitting. So, it is necessary to include pooling layers in the neural network. By doing so, the probability of a reduction in the number of parameters and computation time is increased. The layer applies downsampling by means of function max. It works individualistically on every dimension of the previous layer. Thus, the depth of the next layer is equal to the depth of the previous layer. Similarly, the number of pixels is set when the window is moved, or stride, as the convolutional layer.

There are two types of pooling layers. In traditional pooling, the dimensions of the window are equal to stride. If the dimensions of the window are larger than the stride, then it is referred to as overlapping pooling. Usually, in traditional pooling, the window size is 2x2 and the stride size is 2, and in the overlapping pooling, the window size is 3x3 and the stride size is 2. In addition to max pooling, the other functions are also used: average pooling and L_2-norm pooling. The average pooling computes the window's average to signify the next layer value, and L_2-norm pooling utilizes the L_2-norm. Thus, the deep features are extracted.

12.4.3.2 *Gray Level Co-occurrence Matrix (GLCM)*

GLCM is to describe the texture feature present in gray space. The texture features obtained are contrast, correlation, homogeneity, dissimilarity, entropy, mean and variance. $P_d(i,j)$ represents the joint probability density with the distance (d) between two pixels i and j.

- **Energy**: It is the sum of elements in the image matrix of GLCM after squaring them.

$$Energy = \sum_{i=0}^{L-1}\sum_{j=0}^{L-1} P_d^2(i,j) \tag{12.4}$$

- **Contrast**: It measures the local variations in the matrix of GLCM.

$$Contrast = \sum_{i=0}^{L-1}\sum_{j=0}^{L-1}(i-j)^2 P_d(i,j) \tag{12.5}$$

- **Correlation**: It calculates the joint probability occurrence between the pair of pixels in the matrix of GLCM.

$$Cor = \sum_{i=0}^{L-1} \sum_{j=0}^{L-1} \frac{P_d(i,j)}{1+(i-j)^2} \tag{12.6}$$

- **Homogeneity**: It calculates the nearness of element distribution in the GLCM to its diagonal.

$$Ent = -\sum_{i=0}^{L-1} \sum_{j=0}^{L-1} P_d(i,j) \log P_d(i,j) \tag{12.7}$$

While extracting these features, directions such as horizontal, vertical and 45° angle can be changed. Thus, applying three directional changes for each texture feature given here, we can obtain 12 sets of GLCM features. The obtained GLCM features are as follows:

$$\begin{aligned} \text{GLCM features} = \{ &\text{energyH, energyV, energy45,} \\ &\text{contrastH, contrastV, contrast45,} \\ &\text{correlationH, correlationV, correlation45,} \\ &\text{homogeneityH, homogeneityV, homogeneity45} \} \end{aligned} \tag{12.8}$$

12.4.3.3 Hu Invariant Moments

Let f(x, y) be the distribution based on the gray level, μpq, central moment and ηpq, normalized central moment [27] can be calculated using the equation (12.11) to (12.18):

$$\mu_{pq} = \sum_{x=1}^{M} \sum_{y=1}^{N} (x-x_0)^p (y-y_0)^q f(x,y) \tag{12.9}$$

$$\eta_{pq} = \mu_{pq}/\mu_{00}^r \tag{12.10}$$

Here, $p + q = 2, 3,..., r = (p + q + 2)/2$. As a result, seven features are obtained using invariant moments (Amit Kumar Verma et al. [24]) above and seven-dimensional feature vectors are thus obtained:

$$\varphi_1 = \eta_{20} + \eta_{02} \tag{12.11}$$

$$\varphi_2 = \eta_{20} - \eta_{02}^2 + 4\eta_{11}^2 \tag{12.12}$$

$$\varphi_3 = \eta_{30} - 3\eta_{12}^2 + \eta_{03} + 3\eta_{21}^2 \tag{12.13}$$

$$\varphi_4 = \eta_{30} - \eta_{12}{}^2 + \left(\eta_{21} + \eta_{03}\right)^2 \tag{12.14}$$

$$\begin{aligned}
\varphi_5 &= \left(\eta_{30} - 3\eta_{12}\right)\left(\eta_{30} + \eta_{12}\right)\left(\left(\eta_{30} + \eta_{12}\right)^2\right. \\
&\quad \left. -3\left(\eta_{21} - \eta_{03}\right)^2\right) + \left(3\eta_{21} - \eta_{03}\right) \\
&\quad \times \left(\eta_{21} + \eta_{03}\right)\left(3\left(\eta_{30} + \eta_{03}\right)^2 - \left(\eta_{21} + \eta_{03}\right)^2\right)
\end{aligned} \tag{12.15}$$

$$\varphi_6 = \left(\eta_{20} - \eta_{02}\right)\left(\left(\eta_{30} + \eta_{12}\right)^2 - \left(\eta_{21} + \eta_{03}\right)^2\right) + 4\eta_{11}\left(\eta_{30} + \eta_{12}\right)\left(\eta_{21} + \eta_{03}\right) \tag{12.16}$$

$$\begin{aligned}
\varphi_7 &= \left(3\eta_{21} - \eta_{03}\right)\left(\eta_{30} + \eta_{12}\right)\left(\left(\eta_{30} + \eta_{12}\right)^2\right. \\
&\quad \left. -3\left(\eta_{21} + \eta_{03}\right)^2\right) + \left(\eta_{30} + 3\eta_{12}\right) \\
&\quad \times \left(\eta_{21} + \eta_{03}\right)\left(3\left(\eta_{30} + \eta_{12}\right)^2 - \left(\eta_{21} - \eta_{03}\right)^2\right)
\end{aligned} \tag{12.17}$$

$$F_{H_u} = \left[\varphi_1, \varphi_2, \varphi_3, \varphi_4, \varphi_5, \varphi_6, \varphi_7\right] \tag{12.18}$$

12.4.3.4 Color Moments

The color moments [27] in HSV (Hue, Saturation and Value) space are also utilized. The color moments are calculated using the Eq. (12.19)–(12.21).

$$\mu_i = \frac{1}{N}\sum_{j=1}^{N} P_{i,j} \tag{12.19}$$

$$\sigma_i = \sqrt{\frac{1}{N}\sum_{j=1}^{N}\left(P_{i,j} - \mu_i\right)^2} \tag{12.20}$$

$$s_i = \sqrt[3]{\frac{1}{N}\sum_{j=1}^{N}\left(P_{i,j} - \mu_i\right)^3} \tag{12.21}$$

$P_{i,j}$ means the ith color component of jth pixel in the color image. The color moments are expressed as a vector as shown in Eq. (12.22):

$$F_c = \left[\mu_H, \sigma_H, s_H, \mu_S, \sigma_S, s_S, \mu_V, \sigma_V, s_V\right] \tag{12.22}$$

12.4.4 Feature Selection

The best features have to be selected from the available feature set. To attain the best feature set, autoencoders are used by the proposed system. Autoencoders provide the best way of representing samples in lower dimensions. They were introduced by Geoffrey E Hinton et al. [28]. Also, many other works were proposed using autoencoders [29]. While many other feature reduction techniques may incur loss of information, the feature reduction technique includes only a minimum loss of information. Let R represents the informative representation of its input. Feature selection is based on the following: (1) features paying the most on reconstruction of input using R; (2) selection or rejection of a feature is based on the reconstruction error.

12.4.4.1 Ranking Procedure

Let X be the set of original features, $X = \{F_0, F_1, \dots F_n\}$. The proposed set tries to find the informative subset, Y, consisting of features with more information content. Let $Y = \{G_0, G_1, \dots, G_n\}$. The set Y is now sent for classification at the end of the process. So, when autoencoder is fed with X, it results in an output with Y. It includes two important parts, namely (1) Encoder (W_1) and (2) Decoder (W_2). Here, $G = F \times W_1 \times W_2$.

12.4.4.2 Reconstruction Error (RE) Measure

The reconstruction error value is to find out the importance of every single feature. To find the importance of the ith feature, the value of S_i is computed based on the below Eq. (12.23).

$$S_i = \|A^i(X) - X\|^2 \tag{12.23}$$

The S_i, with low value indicated ignoring the feature will not affect the reconstruction of original data. The values of set S_i are sorted in non-increasing order. The first K number of features are selected, and the remaining is ignored. The value of K is set based on the below Eq. (12.24). (The value of τ is always set to 0.9.)

$$\arg\min_{Z} \left(1 - \frac{\sum_{j=1}^{j=Z} S_j}{\sum_{i=1}^{i=N} S_i} > \tau \right) \tag{12.24}$$

12.4.5 Image Classification

The feature set is then sent for the process of classification. The classification accuracy improves when the input is a fused image [30]. The proposed system uses a very different way of classifying the features and generating a very efficient classified result. Classifying an image using a single classifier may be good sometimes, but it may not be efficient all the time. So, the proposed system tries to combine multiple classifiers and build a great classifier. Many classifiers here are referred to as combined classifier systems (CCS).

12.4.5.1 *Classification Based on BP Algorithm*

The proposed work uses an uncomplicated three-layer neural network. Initially, the labeled samples are used for training. The labeling is done such that the corresponding class gets the value '1' in the class label, and the remaining classes get the value '0' in the class label. For example: If there are five classes, the class label for class 1 will be (1, 0, 0, 0, 0). The class label for class 4 will be (0, 0, 0, 1, 0). The Backpropagation (BP) classifier is trained using training samples and their associated class labels. After the completion of training, the test samples are given, and the output class labels are obtained. The output obtained will be a vector of size 1 × 5. Each column of the output vector represents its similarity to each class through the probability value. The normal activation functions are used as mentioned in Eq. (12.25)

$$a_j^l = \sigma\left(\sum_k w_{jk}^l a_k^{l-1} + b_j^l\right) \tag{12.25}$$

Here, b_j^l and a_j^l represent bias and activation of the lth layer, respectively, where j represents the neuron. Also, w^l represents the weight connecting the neurons of the lth layer of neurons.

12.4.5.2 *Classification Based on k-nearest Neighbor*

The key idea here is to classify a pixel based on the neighborhood values. Consider a training set with (x, y). Now, x_i signifies feature vectors, and $y_i \in Y = \{c_1, c_2, ..., c_m\}$ signifies the classification labels, $i = 1, 2, ..., N$. Given an input x, the $N_k(x)$, k-nearest neighbor distance is computed with training entities. The distance metric used in the proposed system is Euclidean distance. The voting is as shown in Eq. (12.26). The 'y' thus provides the final class label for the given test samples.

$$y = \arg\max_{c_j} \sum_{x_i \in N_k(x)} I\left(y_i = c_j\right) \tag{12.26}$$

Here, I stands for indicator function and is as described in Eq. (12.27):

$$I = \begin{cases} 1, | \ y_i = c_j \\ 0, | \ else \end{cases} \tag{12.27}$$

12.4.5.3 *Classification Based on Naive Bayes*

One of the member classifiers of our proposed system is naive Bayes classifier [31]. Here, the traditional naive Bayes classifier with weight assumption to each attribute is implemented. The training sample is represented as D, where $D = \{x_1, x_2, ..., x_N\}$ with M number of occurrences; each M occurrence has 'n' attribute value with their associated class label. Where, $X = \{x_{i,1}, x_{i,2}, ..., x_{i,n}, y_i\}$ represents the ith occurrence of D. Here, $x_{i,j}$ represents j^{th} attribute and y_i represents the class label. $Y = \{c_1, ..., c_k, ..., c_L\}$ represents category label. In naive Bayes, the model is established using Eq. (12.28):

$$c(x_t) = \arg\max_{c_k \in y} P(c_k) P(x_{t,1}, x_{t,2}, \ldots, x_{t,n} \mid c_k) \tag{12.28}$$

Here, the $P(c_k)$ denotes the class probability, c_k and $P(x_{t,1}, x_{t,2}, \ldots, x_{t,n} \mid c_k)$ denotes joint distribution of x_t conditioned by c_k. we assume that all the variables are independent variables. Thus, the Eq. (12.29) can be rewritten as

$$c(x_t) = \arg\max_{c_k \in y} P(c_k) \prod_{j=1}^{n} P(x_{t,j} \mid c_k) \tag{12.29}$$

Given a test sample x_t, the class label y_t for each x_t can be calculated using Eq. (12.30)

$$c(x_t) = \arg\max_{c_k \in y} P(c_k) \prod_{j=1}^{n} P(x_{t,j} \mid c_k)^{w_j} \tag{12.30}$$

where w_j indicates the attribute j's weight, which is always in the range of [0, 2] and $P(x_{t,j} \mid c_k)$ denotes joint distribution of $x_{t,j}$ conditioned by category c_k.

12.4.5.4 Combined Classifier System (CCS)

The functional diagram for the CCS is shown in Figure 12.3. The proposed system builds CCS with the view that if a different classifier is good at predicting different class labels, then it is diverse. So, an expert classifier is suggested for every class. Let the member classifier be represented as e_i, and Λ_i represent the classification accuracy ranking of e_i class-wise. The L member classifiers contain L rankings. For member classifiers e_i & e_j, calculate their corresponding rankings Λ_i & Λ_j and ranking distance. The calculation based on Spearman distance is done using Eq. (12.31).

$$\rho(\Lambda_i, \Lambda_j) = 2 - \frac{6.\sum_{k=1}^{M} (\Lambda_i(k) - \Lambda_j(k))^2}{M(M^2 - 1)} \tag{12.31}$$

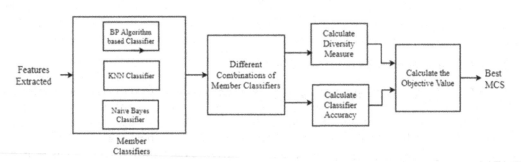

FIGURE 12.3
Functional diagram for the proposed system classification method.

The diversity measure is done between every combination of the member classifiers. The construction of the CCS not only depends on the diversity measure but is also based on the accuracy of the classifier. So, an objective function is proposed to find out the optimal CCS combination of the member classifiers, which could perform well better than the other combination of the classifiers. The objective function is given in the Eq. (12.32)

$$\text{Obj}(\text{CCS}) = w_D.\frac{\text{Div}(\text{CCS})}{2} + w_A.\text{Acc}_{ave}(\text{CCS}) \tag{12.32}$$

Here, $\text{Acc}_{ave}(\text{CCS})$ represents the average member classifier's classification accuracy. w_D and w_A represent the weights associated with the diversity and average accuracy, respectively. Both these are of the user preferences, and the proposed system sets it to value 1. The final best CCS is found using the below formula:

$$\text{CCS}_{best} = \max\left\{\text{Obj}_{(\text{CCS})_1}.\text{Obj}_{(\text{CCS})_1},...,\text{Obj}_{(\text{CCS})_n}\right\} \tag{12.33}$$

The value of n in the above Eq. (12.33) represents the total number of member classifier combinations available.

12.5 Experimental Results and Discussion

12.5.1 Description of Dataset

The experimental results of the system proposed here are proved using the PAN and MS images of Madurai City. The sensor named LISS IV is used for capturing the image, which has a 5.8 m spatial resolution. The spectral resolution of the red band is 0.62 to 0.68 μm, the green band is 0.52 to 0.59 μm, and the near-infrared band is 0.76 to 0.86 μm. The image scale is 1:50000 approximately. The PAN and the LISS IV images are shown in Figure 12.4(a) and (b). The images show the view of a city, Madurai, situated in Tamil Nadu, a state in India. The land cover of the city is approximately 23.5 km × 23.5 km. The city is found with 78 04′ 47″ E to 78 11′ 23″ E longitude and 9 50′ 59″ N to 9 57′ 36″N latitude. The land cover includes mainly urban areas, vegetation, saline land, wasteland and water areas.

12.5.2 Results of the Proposed System

12.5.2.1 Evaluation for Image Fusion

The proposed work carries out the research based on the fusion results only. So, the overall accuracy of the classification will be based on the fusion output. So, evaluating the fusion results plays a vital role. The inputs of the PAN and MS images are given in Figure 12.4 and are used for fusion methodology, and the fusion output is shown in Figure 12.5. The fusion results are evaluated based on specific metrics [32–34] listed below. The evaluation metric of our proposed method and their measures of fusion result are tabulated in Tables 12.2 and 12.3, respectively.

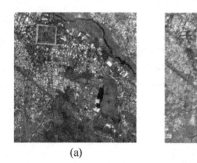

(a)   (b)

FIGURE 12.4
(a) PAN and (b) MS image of LISS IV Madurai City.

FIGURE 12.5
Fused image of LISS IV Madurai City from PAN and MS images.

TABLE 12.2

Image Fusion Evaluation Metrics

S.No	Name of the Evaluation Metric	Explanation	Value
1	PSNR – peak signal-to-noise ratio	Calculates the quality of image.	Higher value is a sign of better quality in image fusion.
2	SA – spectral angle	Shows the resemblance of two vectors $(s,\hat{s})$ based on spectral values.	If the SA value is near zero, then the two images are similar.
3	SSIM – structural similarity index matrix	Represents structural similarity of two images.	Higher SSIM shows more similarity.
4	RMSE – root-mean-square error	Calculates error.	Low RMSE means the fused image has less signal error.
5	CC – correlation coefficient	Specifies the correlation analysis of original & fused images.	If it is near 1, the CC value means both images are alike. Otherwise, it is near 0.
6	FMI – feature mutual information	Shows the image intensity similarity of reference image, MI_{RI} and fused image, $MI_{RI} \, MI_{fI}$.	Higher value denotes better performance.

(Continued)

S.No	Name of the Evaluation Metric	Explanation	Value
7	NCC – normalized cross correlation	NCC value between two images.	Higher value signifies better performance (where '1' takes the maximum value).
8	Gradient	Resemblance between fused and original gradient.	Higher value signifies better performance.
9	SCC – spatial correlation coefficient	Spatial correlation between fused and reference image.	Higher value signifies better performance.
10	Image entropy	Fused image's information content.	Higher value signifies high information content in the fused image.

TABLE 12.3

Image Fusion Accuracy Measures

Metrics	Values of the Metrics	Metrics	Values of the Metrics
PSNR	56.1	RMSE	50.68
SA	0.4	FMI	1.4
SSIM	1.3	Gradient	0.956
NCC	0.986	SCC	0.986
CC	0.987	IE	0.987

TABLE 12.4

CCS Classification Accuracy and Fusion-based Accuracy on LISS IV Dataset

Members in CCS	Ranking Distance Div(CCS)/2	Average Accuracy Accave(CCS)	Obj(CCS)
{1,2}	0.75	0.55	1.30
{1,3}	0.60	0.73	1.33
{2,3}	0.50	0.73	1.23
{1,2,3}	0.75	0.73	1.48

12.5.2.2 Evaluation for Classification

The three different classifiers are given here. The CCS tries to find the best combinations of features. So, the ranking distance based on Eq. (12.32) and objective function value as shown in Eq. (12.33) is calculated and tabulated in Table 12.4. As the objective function value increases, the classification accuracy is improved and vice versa. The {1,2,3} combination of classifier combination performs well and can be employed in our method. The accuracy of the classifier is computed by means of the accuracy metrics as mentioned in Table 12.5 [35].

The overall classification accuracy metrics are listed in Table 12.5 and based on which the classified image is evaluated. The classification accuracies for the proposed work before and after fusion are tabulated in Tables 12.6 and 12.7. The classification accuracies for

TABLE 12.5

Classification Accuracy Metrics

S.No	Name of Evaluation Metric	Explanation
1	UA – user accuracy	Represents the proportion of pixels that are correctly classified and the total quantity of pixels in the specific class in the confusion matrix.
2	PA – producer accuracy	Represents the proportion of correctly classified pixels in a particular category to the total quantity of pixels utilized for training in that class.
3	KC – kappa coefficient	Based on a statistical test. The value ranges between −1 to +1.
4	Overall accuracy	Represents the proportion of correctly classified pixels compared to all reference pixels.

different classifiers and the combination of member classifiers are tabulated similarly. The output LISS IV image for the classification is shown in Figure 12.5.

12.5.3 Discussions about the Proposed System

The proposed system took a different try of utilizing the image fusion results as an input to the classification system. Unlike earlier methods for fusion of LISS IV images, the proposed system uses the deep convolution-based neural network for fusing the image as yet another turning point. The CNN with the specifications given in Table 12.1 is used. The image fusion thus produces an image whose clarity is improved and is obvious. Tables 5.5 and 5.6 show that the image classification with image fusion outperforms classification without fusion with an overall accuracy of 92.3 percent. Also, it is clear from Figure 12.6.

The fused image is sent for feature extraction as input, and the features of the image are extracted. The extracted features are of larger quantity, and the involvement of autoencoders has played a significant role in selecting the robust features. The features that can contribute much to the classification process are identified and selected. Another important point with the proposed system is the classification system used. The traditional classification method with one classifier is replaced with the combined classifier system. Thus, combining the different classifiers still improves the proposed system's classification accuracy. From the results and the analysis of the whole work, we conclude that we have found a better method that could outperform many other classification techniques for RS images.

TABLE 12.6

Classification Accuracies for Different Individual Classifiers (Without Fusion)

| Classifier | Without Fusion | | | |
	UA	PA	KC	Overall Accuracy
BP	**88.9**	86.2	0.81	87.1
KNN	84.0	80.3	0.79	**93.3**
Naïve Bayes	85.1	86.2	0.78	85
CCS	88.6	**88.4**	**0.85**	89.1

TABLE 12.7

Classification Accuracies for Different Individual Classifiers (With Fusion)

Classifier	With Fusion			
	UA	PA	UA	Overall Accuracy
BP	92.9	91.2	**0.88**	92
KNN	89.0	79.0	0.83	91.3
Naïve Bayes	93.1	91	0.87	90
CCS	**94.6**	**96.4**	**0.88**	92.3

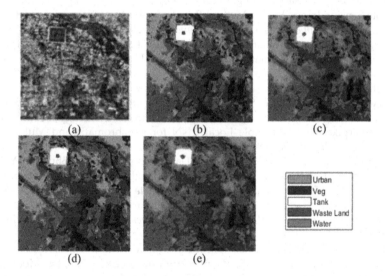

FIGURE 12.6
Output results of proposed work for LISS IV images: (a) fused image; (b) output image results of BP classifier; (c) output image results of NN classifier; (d) output image results of naive Bayes; (e) output image results of CCS classifier.

12.6 Conclusion

Our proposed system explains the effectiveness of image fusion techniques in image classification. As the LISS IV image contains many details, exploring it needs an enhanced input image. Thus, the objective is satisfied through the proposed method. The experimental results are done to still prove its effectiveness. The fusion results are evaluated through many metrics, and the classification results are measured through many classification metrics. The feature extraction and the feature selection are also performed efficiently. The evaluation parameters are found to be more satisfying than expected. The proposed system outperforms many existing methods. In the future, the proposed system can also be tested with other land cover areas sensed by different satellites.

References

1. Jinying Zhong, Bin Yang, Guoyu Huang, Fei Zhong, Zhongze Chen (2016), "Remote Sensing Image Fusion with Convolutional Neural Network," *Sensing and Imaging*, Vol. 17(10), pp. 1–16.
2. Yong Yang, Lei Wu, Shuying Huang, Weiguo Wan, Yue Que (2018), "Remote Sensing Image Fusion based on Adaptively Weighted Joint Detail Injection," *IEEE Access*, Vol. 6, pp. 6849–9864.
3. Shashidhar Sonnad (2016), "A Survey on Fusion of Multispectral and Panchromatic Images for High Spatial and Spectral Information," *International Conference on Wireless Communications, Signal Processing and Networking (WiSPNET)*, pp. 177–180.
4. Yong Yang, Weiguo Wan, Shuying Huang, Feiniu Yuan, Shouyuan Yang, Yue Que (2016), "Remote Sensing Image Fusion Based on Adaptive IHS and Multiscale Guided Filter," *IEEE Access*, Vol. 4, pp. 4573–4582.
5. Chang Han, Hongyan Zhang, Changxin Gao, Cheng Jiang, Nong Sang, Liangpei Zhang, (2016), "A Remote Sensing Image Fusion Method Based on the Analysis Sparse Model," *IEEE Journal of Selected Topics in Applied Earth Observations and Remote Sensing*, Vol. 9(1), pp. 439–453.
6. Kai Zhang, Min Wang, Shuyuan Yang, Licheng Jiao (2018), "Convolution Structure Sparse Coding for Fusion of Panchromatic and Multispectral Images," *IEEE Transactions on Geoscience and Remote Sensing*, Vol. 57(2), pp. 1117–1130.
7. Mehdi Ghamchili, Hassan Ghassemian, (2017), "Fusion of remote sensing images based on dictionary learning," *Iranian Conference on Electrical Engineering (ICEE)*, pp. 1895–1900.
8. Wei Zhao, Licheng Jiao, Wenping Ma, Jiaqi Zhao, Jin Zhao, Hongying Liu, Xianghai Cao (2017), "Superpixel-Based Multiple Local CNN for Panchromatic and Multispectral Image Classification," *IEEE Transactions on Geoscience and Remote Sensing*, Vol. 55(7), pp. 4141–4156.
9. Xu Liu, Licheng Jiao, Jiaqi Zhao, Jin Zhao, Dan Zhang, Fang Liu, Shuyuan Yang, Xu Tang (2017), "Deep Multiple Instance Learning-Based Spatial–Spectral Classification for PAN and MS Imagery," *IEEE Transactions on Geoscience and Remote Sensing*, Vol. 56(1), pp. 461–473.
10. Chetna Soni, Manoj Joseph, A. T. Jeyaseelan, J.R. Sharma (2017), "Automatic Extraction of Built-Up from SAR Imagery," *IEEE International Conference on Computational Intelligence and Computing Research (ICCIC)*.
11. Dileep Kumar Gupta, Rajendra Prasad, Pradeep Kumar, Varun Narayan Mishra, Ajeet Kumar Vishwakarma, Ravi Shankar Singh, Vinayak Srivastava (2015), "Spatial Modeling of SPAD Values for Different Type of Crops Using LISS-IV Satellite Imagery," *International Conference on Microwave, Optical and Communication Engineering (ICMOCE)*, pp. 212–215.
12. Pradeep Kumar, Rajendra Prasad, Varun Narayan Mishra et al. (2015), "Artificial Neural Network with Different Learning Parameters for Crop Classification Using Multispectral Datasets," *International Conference on Microwave, Optical and Communication Engineering (ICMOCE)*, pp. 204–207.
13. Varun Narayan Mishra, Rajendra Prasad, Pradeep Kumar, Dileep Kumar Gupta, et al. (2016), "Slope Stability Analysis in a Part of East Sikkim, Using Remote Sensing & GIS," *2nd International Conference on Next Generation Computing Technologies (NGCT)*, pp. 51–60.
14. Gyanesh Chander, Michael J. Coan, Pasquale L. Scaramuzza (2008), "Evaluation and Comparison of the IRS-P6 and the Landsat Sensors," *IEEE Transactions on Geoscience and Remote Sensing*, Vol. 46(1), pp. 209–221.
15. C. R. Prakash, Mahbooba Asra, J. Venkatesh, B. Sreedevi (2015), "Monitoring Urban Land-Cover Features using Resourcesat LISS-III Data," *International Journal of Advanced Remote Sensing and GIS*, Vol. 4(1), pp. 1064–1069.
16. T. Ganesh Kumar, D. Murugan, K. Rajalakshmi, T. I. Manish (2015), "Image Enhancement and Performance Evaluation Using Various Filters for IRS-P6 Satellite LISS IV Remotely Sensed Data," *Geofizika*, Vol. 32(2), pp. 179–189.
17. S. Arivazhagan, J. Praislin Anisha (2013), "Image Fusion Using Spatial Unmixing, International Conference on Signal Processing," *Image Processing & Pattern Recognition*.

18. Shardha Porwal, Sunil Kumar Katiyar (2014), "Performance Evaluation of Various Resampling Techniques on IRS Imagery," *Seventh International Conference on Contemporary Computing (IC3).*

19. S. Rajesh, T. G. Nisia, S. Arivazhagan et al. (2020), "Land Cover/Land Use Mapping of LISS IV Imagery Using Object-Based Convolutional Neural Network with Deep Features," *Journal of Indian Society of Remote Sensing,* Vol: 48, pp. 145–154.

20. S. Rajesh, S. Arivazhagan, K. Pratheep Moses, R. Abisekaraj (2012), "Land Cover/Land Use Mapping Using Different Wavelet Packet Transforms for LISS IV Madurai Imagery," *Journal of the Indian Society of Remote Sensing,* Vol. 40(2), pp. 313–324.

21. S. Arivazhagan, L. Ganesan (2003), "Texture Classification using Wavelet Transform," *Pattern Recognition Letters,* Vol. 24, pp. 1513–1521.

22. S. Rajesh, S. Arivazhagan, K. Pratheep Moses, R. Abisekaraj (2013), "Genetic Algorithm-Based Feature Subset Selection for Land Cover/Land Use Mapping Using Wavelet Packet Transforms," *Journal of the Indian Society of Remote Sensing,* Vol. 41(2), pp. 237–248.

23. S. Rajesh, S. Arivazhagan, K. Pratheep Moses, R. Abisekaraj (2014), "ANFIS Based Land Cover/ Land Use Mapping of LISS IV Imagery Using Optimized Wavelet Packet Features," *Journal of the Indian Society of Remote Sensing,* Vol. 42(2), pp. 267–277.

24. Amit Kumar Verma, P. K. Garg, K. S. HariPrasad, V. K. Dadhwal (2016), "Classification of LISS IV Imagery using Decision Tree Methods," *The International Archives of the Photogrammetry, Remote Sensing and Spatial Information Sciences,* Vol. XLI-B8, pp. 1061–1066.

25. P. S. Prakash, K. D. Soumya, H. A. Bharath (2018), "Urban building extraction using satellite imagery through Machine Learning," *IEEE Symposium Series on Computational Intelligence (SSCI),* pp. 1670–1675.

26. Zhenfeng Shao, Jiajun Cai (2018), "Remote Sensing Image Fusion with Deep Convolutional Neural Network," *IEEE Journal of Selected Topics in Applied Earth Observations and Remote Sensing,* Vol. 11(5), pp. 1656–1669.

27. Jinhua Liu, Deqiang Han, Yi Yang (2018), "Pest Identification Based on Multiple Classifier System," *37th Chinese Control Conference (CCC),* pp. 9535–9539.

28. Geoffrey E. Hinton, Ruslan R. Salakhutdinov (2006), "Reducing the Dimensionality of Data with Neural Networks," *Science,* Vol. 313(5786), pp. 504–507.

29. Sasan Sharifipour, Hossein Fayyazi, Mohammad Sabokrou, Ehsan Adeli (2019), "Unsupervised Feature Ranking and Selection Based on Autoencoders," *IEEE International Conference on Acoustics, Speech and Signal Processing (ICASSP),* pp. 3172–3176.

30. Lalit Kumar, Priyakant Sinha, Subhashni Taylor (2014), "Improving Image Classification in a Complex Wetland through Image Fusion Techniques," *Journal of Applied Remote Sensing,* Vol. 8, pp. 1–17.

31. Juan Yang, Zhiwei Ye, Xu Zhang, Wei Liu, Huazhong Jin (2017), "Attribute Weighted Naive Bayes for Remote Sensing Image Classification Based on Cuckoo Search Algorithm," *International Conference on Security, Pattern Analysis, and Cybernetics (SPAC),* pp. 169–174.

32. Mustafa Oral, Sultan Sevgi Turgut (2018), "A Comparative Study for Image Fusion," *Innovations in Intelligent Systems and Applications Conference (ASYU).*

33. K. Rajarshi, Ch. Himabindu (2016), "DWT Based Medical Image Fusion with Maximum Local Extrema," *International Conference on Computer Communication and Informatics (ICCCI).*

34. M. Vani, S. Saravanakumar (2015), "Multi Focus and Multi Modal Image Fusion Using Wavelet Transform," *3rd International Conference on Signal Processing, Communication and Networking (ICSCN).*

35. Varun Narayan Mishra, Rajendra Prasad, Pradeep Kumar, Dileep Kumar Gupta et al. (2015), "Evaluating the Effects of Spatial Resolution on Land Use and Land Cover Classification Accuracy," *International Conference on Microwave, Optical and Communication Engineering (ICMOCE),* pp. 208–211.

13

Human Behavioral Identifiers: A Detailed Discussion

T. Suba Nachiar
Velammal College of Engineering & Technology Madurai, India

T. Shanmuga Priya
Associate Consultant at Vuram, India

P.R. Hemalatha
Velammal College of Engineering & Technology, Madurai, India

J.V. Anchitaalagammai
Velammal College of Engineering & Technology, Madurai, India

CONTENTS

DOI: 10.1201/9781003206736-13

13.1 Introduction to Biometric Technology

The word means bio – life, and metric – measure, originating from the Greek language. Biometrics is a technology to ensure a person's identity based on their physiological characteristics. This methodology is fully performed by the system. Forensic techniques, like DNA, latent thumbprints, and hair, are not considered part of this field. This automated identification technology can be executed on fruits, vegetables and animal-produced goods, and can be imposed only on living beings. Depending on the characteristic behavior, this is called "anthropometric authentication."

Statistical techniques mainly use the thumbprint impression, which is used to distinguish or link a group of individuals, but biometrics focuses on identifying the person. It generally gathers data on mental and physical behavior which mainly diverges or is pretty common over a population. Usually, the technologies are only one or the other, though some traits seem more socially or physiologically inclined. The behavioral factor of all biometric actions equates with the human factor aspect to biometric authentication as well.

"Biometric" is a shortened form of the term "biometric authentication"; the latter authentication has been widely used in the field of biology to examine data statistically and quantitatively. So this term here refers to identifying a person with the help of the system. Despite the similarities and variations of the person from the population, maintaining an authentic identity is tedious, irrespective of any biometric technology.

This methodology can be linked to a biometric design and data on the mental and personal characteristics of a person, which are fed into the system at the time of initiation. These structures fundamentally require no identity data and thus allow unidentified recognition.

The system's performance mainly depends upon the communication of persons with the automated mechanism. The interface of technology with human functioning and thinking makes them (biometrics) a more striking topic.

13.2 Historical Outline

The primary sign of an identification (biometric) system was in France in the 1800s. Alphonse Bertillon was the one who established a technique of precise physical dimensions for classifying and comparing criminals. This one was far from perfect, but it paves the way to use exclusive biological features to validate any individual.

Fingerprinting followed suit in the 1880s, which was not only used to classify criminals but also used as a form of signature on agreements. This was a representative of an individual and could be held responsible for it. Edward Henry is credited with the expansion of the thumb impression method called the Henry Classification System, though there has been discussion as to who developed it for identification purposes.

This stood as the initial and foremost scheme, used to identify an individual grounded on the exclusive constructions of impressions. Law enforcement adopted this method, replacing Bertillon's method. It then became standard for criminal identification. This commenced an era's value of study on what other discrete biological features might be used for identification.

13.3 The Basic Characteristics

The primary attributes used for the identification system include physical and behavioral characteristics, such as speech, hand, facial features, impressions, etc. This list still grows to light reflectance on the skin, the odor of the body, and so on. Due to the large variety of fields present, the imaging requirements also vary. The machine can measure single one-dimensional waves signal, i.e., speech; synchronized one-dimensional signals, i.e., writing; a solo two-dimensional image similar to impression; multi-two-dimensional values, such as hand geometry; a time series of two-dimensional visuals, i.e., face and retina; or a three-dimensional visual, i.e., facial identification system.

The physical or behavioral traits of a human being are regarded as a biometric reorganization of an individual, and this should satisfy the basic characteristics like collectability, circumvention, distinctiveness, performance, and universality. Every individual will surely possess a biometric attribute. Let's consider an example: almost everyone in the world will have minimum face-to-face proof of identification, but gait-based identification will be difficult for wheelchair users.

13.3.1 Collectability

Biometric features have been measured quantitively. These values may sometimes be vital in certain applications. Gathering and analyzing the impression in biometrics may be seen as easy, while the same for DNA is difficult in one such example. DNA identification may not be perfect for the E-passport application. But the day-to-day biometric machine might focus on other issues, such as performance. To determine the performance of a biometric system, the speed (i.e., throughput), recognition accuracy, resource necessities and strength for operative and accurate ecological aspects are the vital features. For instance, fingerprint biometrics are small and compact; meanwhile, DNA biometrics slows down, is expensive and works severely.

13.3.2 Circumvention

Generally, the identification system can be evaded by fake approaches. This practical method must be fast with identification accuracy with strong resource necessities, safe for the users, acceptable to all people and appropriately robust to numerous fake approaches. A wide variety of biometric attributes are used in various applications, each with its advantages and disadvantages.

13.3.3 Distinctiveness

Every individual is exclusive with their own biometric attributes. This is mainly why certain biometrics differentiate individual fingerprints. DNA is the most effective method to provide accurate results.

13.4 Biometric Types

The identification process in biometrics is carried out with the exclusive feature of an individual's parameter compared with the existing database. Biometric readers are mainly used to retrieve this personal data. The parameter identified is one of the basic human features. This mainly comes under either physical or behavior identifier [5]. The physical identifiers are the static physical characteristics of an individual. This includes those that are immutable and device-independent.

13.4.1 Fingerprints

This biometric technology enables the users to access the services using their fingerprint images. Scanning for fingerprints has become common over the years, mainly due to mobile phones. Most touch-screen devices (touchpad, mouse, mobile screen, iPad, door lock) have become easy to use with fingerprint detection.

13.4.2 Photo and Video

This type of biometric technology needs a device with a camera for effective verification. Face identification and iris scans are two widely known basic methods. Other image-based authentication techniques are ear recognition, hand geometry recognition, retinal scanning, and vein identification.

13.4.3 Speech

In this technique, specialized software and systems are created to detect, distinguish and authenticate a specific speaker's voice through telephone-based service portals. Voice-based digital assistants are already aware of voice recognition patterns and use this technique to recognize users and authenticate clients.

13.4.4 Signature

This technique authenticates the individual by measuring their handwritten signatures. Digital signature scanners are commonly used by banks and in retail checkouts. They are suitable options for situations where clients or users are expected to sign virtually (e-sign).

13.4.5 DNA

DNA (deoxyribonucleic acid) matching. Nowadays, DNA matching technology is practiced largely in law execution to recognize suspects.

13.5 Behavioral Identifiers

These identifiers are to analyze the individual's characteristic behavior – those characteristics innate in every process of the action produced. Currently, this biometric style is frequently used to distinguish between humans and machines. This contributes to companies to identify spam or to sense cyberattacks. As the science evolves, machines improve in identifying the person with high accuracy, but it sometimes gets reduced in effectiveness in segregating people and machines. Here are some common approaches.

13.5.1 Inputting Forms

Every individual has a unique writing (typing) style. The uniqueness can be identified and measured by their entering swiftness, the time consumed to change each letter, the force of impact on the keypad, etc.

13.5.2 Physical Movements

The walking style of every individual is unique, and these characteristics can also be implemented to identify a person in a structure (company building). This sometimes acts as a subordinate coating to identify the individual for delicate places.

13.5.3 Navigation Forms

Finger and mouse actions on touchpads, screens or mouse pads are exclusive to a person and seem moderately easy to sense in the system. There is no extra hardware needed for this identification method.

13.5.4 Engagement Patterns

Every human intermingles with systems differently, as in opening or handling apps, the average battery we maintain, the way we navigate websites, places with the duration of time of day in which we mainly use our device, how often we check our social media accounts, in what manner we incline the phones while using them or even how we grip the phones are all taken into account as unique attributes of an individual. These characteristics can be used to separate humans from robots, till the robots become accurate in replicating human behavior. And this can also be used as an identification method or, if it gets better enough, as a unilateral security feature.

13.6 Applications

The biometric application comprises its function, operation and part of its methodology. Biometric technologies function in a wide range of applications that vary in terms of performance requirements, operational environment and privacy impression. On implementation, which modality is to be utilized and what hardware and software to deploy is typically driven in large part by the application. The following domain areas use biometric solutions to meet their security needs

13.6.1 Financial Sector

Popular secure application systems such as financial identification, verification, and authentication in commerce help make banking, purchasing, and account management safer and more convenient and responsible for the client or users. In the financial sector, biometric solutions help ensure that a customer is the authentic person they are claiming to be when accessing sensitive financial data by entering their unique biometric characteristics and comparing them to a model stored in a device or on a server. Banking solutions and payment technologies currently use a wide range of biometric modalities, like face, fingerprint, voice, palmprint, hand geometry, iris, retina scan, voice, DNA, signatures, gait, keystroke and other types of biometric recognition, which are all used alone or combined in a multifactorial manner as a system, to lock accounts and counter fraudulent activities.

13.6.2 Security

Network connectivity allows various parts of the world to connect to one another. At this juncture, traditional security methods are not very capable of protecting the most important ones (data, function, operation, etc.). Luckily, biometric technology is more accessible than ever and ready to provide additional security and convenience for everything that needs to be protected, i.e., from a phone's PIN to a car door.

13.6.3 Mobile Application Domain

The mobile biometric techniques are conscious of the intersection of connectivity and identity. It combines one or more biometric attributes for the authentication process and benefits tablets, smartphones, other types of handhelds, wearable technology, and the Internet of Things for adaptable deployment capabilities. The adaptability brought by modern mobile technology, as well as the proliferation of mobile paradigms in the consumer, public, and private world, mobile biometrics makes them more important.

13.6.4 Justice, Law and Enforcement Applications

Law enforcement and biometric technology have a long history, and many significant innovations in identity management have emerged from this beneficial relationship. The biometrics applied by the police force is genuinely multimodal. Face, fingerprint and voice recognition plays a unique role in improving public safety and keeping track of people.

13.6.5 Public Services Applications

13.6.5.1 Healthcare

In the field of healthcare, biometrics introduces an enhanced model. Medical records are among the most valuable personal documents; doctors need to be able to access them quickly, and they need to be accurate. A lack of security and good accounting can make the difference between timely and accurate diagnosis and health fraud.

13.6.5.2 Border Control and Airports

A key area of application for biometric technology is at the border. Biometric technology helps to automate the process of border crossing. Reliable and automated passenger screening initiatives and substation automation systems (SAS) help facilitate international passengers' travel experience while improving the efficiency of government agencies and keeping borders safer than ever before.

13.6.6 Eye Movement Tracking Applications

13.6.6.1 Aviation

Flight simulators track the pilot's eye and head movements in order to analyze the pilot's behavior under realistic circumstances. This simulator can evaluate a pilot's performance based on his eye movements combined with other information. It can also be used as an important training tool for new pilots in order to help them to look at the primary flight display (PFD) more regularly in order to monitor different airplane indicators.

13.6.6.2 Automotive Industry

There is an established relationship between eye movement and attention. Thus, tracking a car driver's eye movements can be very helpful in measuring the degree of sleepiness, tiredness, or drowsiness. The driver's sleepiness can be detected by analyzing either blink duration and amplitude or the level of gaze activity.

13.6.6.3 Screen Navigation

One of the most important applications for people with disabilities is screen navigation. Using cameras, the application can track a person's eye movements in order to scroll a web page, write text, or perform actions by clicking on buttons on a computer or mobile device. This kind of application has been gaining more attention recently, due to the rapid development and growing need for new means of screen navigation, especially on mobile device platforms.

13.7 The Rise of Static Biometric Authentication through Physical Characteristics

The usage of static biometrics or identifiers has increased over the years. Companies and organizations using this have raised serious concerns about the use of physical factors.

- Using only one physical biometric data point to authenticate a user at the time of login is fundamentally the same as adding a static second password, albeit one that can never be changed if compromised.
- Physical biometrics can be captured and are sold in many cases, utilized again or synthesized with fake IDs.
- This creates several issues, which have forced many large organizations like IBM to withdraw or scale back from facial recognition technologies.

Physical biometrics are purely based on a static approach. The problem with static biometrics security based on several factors, like points captured in fixed images, is that even if the initial authentication is valid and done by the legitimate user, the integrity of the session gradually erodes over time. The only mode to restore it is to require additional authentication factors. But, continuing to ask users or clients for traditional attributes – passwords, facial recognition, fingerprints – is troublesome and causes significant resistance, leading to poor client experience.

In today's digital era, clients or users not only expect, they urge, to be able to access their user accounts and their operations as seamlessly as possible. Nowadays, companies stay competitive and improve the digital experience while still keeping customers secure.

13.8 Behavioral Biometrics in Today's Digital World

Behavioral identifiers are dynamic ones that raise trust values, reduce friction during online interactions and lead to the detection of fraudulent activities. These identifiers run in the background and protect interactive sessions post-login; they detect indirect anomalies based on risk.

Behavioral identifiers capture the individual's physical and cognitive digital behavior as their input. They classify the individual by their activity. They categorize behavior patterns of legitimate users vs. illegitimate human or non-human cybercriminal actors. They provide a solution to cybercrime. The identifiers detect fraud in an age when offenders have more access to personally identifiable information (PII) and more hacking methodologies than ever before.

Our dependence on digital services is increasing day by day, which makes cyber attackers more comfortable. Most applications use or allow third-party providers to connect to their core systems through Application Programming Interface (APIs). Through this, the customers can connect directly from their accounts. Many of the security features and fraud control measures companies have in place will still not be able to stop fraudulent activities from attacking accounts via third-party providers. These solutions are not set up to monitor sessions originating with third-party providers (TPPs.)

Risk factors start at the account opening process. Here, the customers will open an account with a TPP, and the company or institution will let the TPP link to the user's account. Fraudsters are trained to use synthetic identities to open fake accounts.

If TPPs fail to encounter fraudulent activities in the account opening process, then the companies or the institutions should find a way to identify fake accounts, to protect themselves and their customers. Even the TPPs do have to encounter certain fraudulent activities and be able to detect them to stay in operation. Various security measures have to be imposed to become smarter both to detect new avatars of cybercrime and to support the growth of the digital era. Behavioral identifiers provide an effective solution to achieve the standard level of security. Institutions use behavioral biometrics to measure the rate of fraud in their TPPs to make sure they are behaving as secure partners. This is possible through continuous monitoring of user behavior before and after login to differentiate between fraudsters and legitimate users.

Behavioral identifiers capture and use the customer's cognitive and physiological digital behavior to differentiate authentic users from criminals or attackers. This technique is practiced to trace fraudulent activities, identify theft and provide a good experience to the customers. This is achieved through profiling customer behaviors such as typing cadence, swipe patterns, and mouse movements. Later, the activity of the customer or user is compared and associated against the primary user profile for the individual account to provide a passive authentication layer and against population-level patterns to identify statistically observed norms for legal and illegal behavior. Usually, cyber attackers input the data in an entirely different way from authentic users.

Cyber attackers repeatedly delete and fix errors. They do not present familiarity with the data; the attackers rely heavily on copying and pasting, particularly in automated programs. Attackers are much more accustomed to new accounts, due to multiple application forms, while authentic users rely on the auto-fill feature for personal details. Each one possesses a unique pace and navigation pattern.

13.9 Analyzing the Patterns in Human Activity

Behavioral biometrics is the science of identifying individuals based on their behavioral and biological characteristics. There are three main categories in the development of behavioral biometrics: kinesthetics, device-based gestures, and vocal patterns.

13.9.1 Physical Movements

- Gait
- Handling
- Posture

In the active state, individual weight distribution data can be collected. Gait recognition emphasizes a special way for an individual to make movements that add up to a particular style. The individual's movement speed, posture, and stride length are considered. The device handling is also taken into account, i.e., how the individual holds it, handles it and so on.

13.9.2 Voice Biometrics

Massive advances in neural networks over the past few years have enhanced the development of biometric voice algorithms that are accurate, faster and can identify users with a small amount of input. Some have even used this subject in their fictional writing. Behavioral biometrics identifies a vocal pattern of an individual based on sound variations that are most common in a human's speech. It improves the customer experience in login processes and minimizes the frustration that occurs due to lost and stolen credentials [3].

13.9.3 Device-based Gestures

A touch-based unified user authentication mechanism supports both passive and continuous authentication of the users based on their touch gestures. Individuals disclose their unique touch features, like the speed and acceleration of movement, finger pressure and trajectory upon the device. How the individual handles their device also leaves an imprint on it, like mobile interactions and keystroke dynamics. Typing patterns differ for every individual and this feature was used to recognize telegraph operators in the early 1990s. These unique features allow us to trace keystroke dynamics, such as the speed of typing, duration of keystrokes, etc. Along with mobile interactions, individual habits of correcting mistakes or errors made in the text have also been taken into account for identifying an individual. Even the usage of the cursor, speed, clicks, paths, and direction changes are analyzed and taken into account to authenticate a person.

13.10 Emerging Technologies in Behavioral Biometrics

13.10.1 Human Behavioral Patterns

Patterns of behavior can be decomposed into semi-behaviors. The humanoid eye cannot differentiate them, but the software picks them up to create a profile. Movement patterns comprise several vital features which are drafted on an individual's unique characteristics. It may seem that the individual is no different from the crowd, but certainly their social habits, a unique way of typing words in a language that is not native to the individual, in combination, add up to a recognizable interaction with the user device.

13.10.2 Sensors

Sensors are small but highly efficient. A biometric sensor is used for identification and authentication purposes. These biometric sensor devices use automated methods for verifying the identity of an individual based on a physical attribute, including fingerprints, facial images, iris, and voice recognition, etc., Almost every modern personal electronics product, from mobile devices and wearables to home appliances, are embedded with sensors that can be configured in a way to collect data passively. The accelerometer and a gyroscope are suitable for these purposes, and they are found in almost every mobile device. Behavioral biometrics focuses not so much on the outcome of individual actions as on the way the individual performs those actions. Analysis of the received data can help in the identification and authentication of an individual.

13.11 Machine Learning/Deep Learning

Machine learning/deep learning is capable of learning from human behavior and enhances the user profile that helps in authenticating the individual or transactions. It can compete with human perception and even surpass it. This can be accomplished through the development of algorithms in software that collect and analyze the data of the individual.

13.11.1 How it Works

Recent fraud controls often treat clients or users like offenders, introducing additional friction into the user experience. This usually happens in the online account opening process, where the customer applications are deferred for manual review, which can incur high operational costs. Behavioral biometrics helps in detecting account opening fraud and account takeover (ATO) by understanding behavioral intent to identify illegal activity versus that of a legal applicant. False declines of applications and transactions are controlled by analyzing user digital behaviors to assess the risk of an activity and driving the appropriate action. Usually, the model is designed with the customer experience in mind. It is invisible to the individual user. It allows the consumers to go about their banking activities while also being guaranteed utmost security. Placing the tools in the appropriate order will give the customer an excellent experience and provide a balance between trust and risk that is properly calculated and aligned to business priorities.

Continuous monitoring provides constant and stable protection. It is considered to be advanced to traditional forms of (login-based) authentication because, while login-based authentication checks a user's identity only once, at the start of a login session, continuous authentication recognizes the correct user for the duration of ongoing work. So, it can spot the moment an illicit attacker seizes control of the session, immediately ending the session, logging the account out, and protecting critical systems and data. Thus, it reduces the deception losses and builds trust in digital interactions. To achieve stable protection, continuous collection and analysis of data throughout the session are essential, so even the most subtly changing situations within the session do not go undetected. They are driven by ML algorithms that analyze the cognitive digital and physiological behavior of individuals across web and mobile channels. Every individual behaves in a completely distinct way. The fluctuations in vocal tone as they speak, and the cadence with which they type are as ideal as fingerprints—but are much harder for malicious players to capture, much less duplicate. Behavioral biometrics uses these types of patterns to authenticate the individual and protect their data. It considers the real-time physical interactions such as keystrokes, mouse movements, swipes, and taps.

Usually, behavioral biometric tools run on mobile devices and systems connected to an organization's data. Every user with valid access automatically generates a behavioral profile that reflects the unique ways in which they interact with critical systems.

Once a user's profile is learned, their gestures are monitored silently, in real time, to continuously authenticate identity. The user's digital behavior on various levels, like behavioral identification, cognitive analysis and behavioral insights, begins to be analyzed.

1. Comparing the current sessions with the distinct user profiles to identify the anomalies like human versus robot, chatbot activities, etc.

2. User profiling at the population level is done to recognize or detect behavior patterns of authentic users and offenders (attackers).

3. Combining individual and population-level profiles to determine user intent and emotional state in the context of activity to detect complex situations indicating high levels of risk.

4. Now, it analyzes every user session and generates a risk score based on this deep user behavioral profiling. Depending on the risk score, organizations can initiate supplementary actions like requiring step-up authentication or manual review. It also provides organizations with the top threat indicators to allow further visibility into risk. Confirmed fraud feedback is incorporated to continually enhance the model's accuracy and adapt to new and emerging attacks. With over a decade of experience analyzing user interaction data, it sometimes offers out-of-the-box risk models that clients can leverage to gain actionable insights into fraudulent activities and realize immediate value upon deployment.

If behavioral patterns that do not match the profile occur, the system can immediately prompt for other forms of authentication, block access, or lock the device down entirely.

13.12 Behavioral Biometrics Examples

Behavioral biometrics is currently used in internet banking, payments, e-commerce, and high-security authentication markets. Over a period of years, the use of behavioral identifiers by companies has grown.

13.12.1 Compromised Credentials

Login credentials can be stolen or compromised, regardless of the countermeasures taken. A behavioral identifier can help validate the granted access to ensure system security.

13.12.2 Account Details/Password Sharing

Dissemination of information about accounts is definitely risky. The above-defined technologies can differentiate authentic users, and attackers can determine when an authentication block is needed.

13.12.3 User Substitution

Substitution for one individual may happen in some cases, especially in areas that use outsourcing. To avoid such risks, in unusual inconsistent substitution, the behavioral identifiers will help to determine the client or user authentication.

13.12.4 Remote Access Trojans

If unauthorized individuals try to access a system, the behavioral identifier can discriminate such subjects from authorized ones by their collected biometric profiles. This is the only possible way, regardless of whether remote access is being used or not.

13.12.5 Insider Threats

The emergence of an internal threat is possible when the access of some users to the system leads to the opening of the door to access by other users, even if it happened unintentionally. Behavioral biometrics allows you to determine in real time that the right person is using the system.

13.12.6 USB Rubber Ducky Attacks

This attack is quite common and is based on the usage of a mouse or keyboard, while this one is entered automatically. Various algorithms have been developed that can detect this type of threat. It also has the ability to block further input.

13.12.7 Phishing Attacks

While technology has yet to keep people from clicking on suspicious links, behavioral identifiers know how to deal with the consequences. It recognizes cybercriminals who have already obtained the data they need to log in.

13.12.8 Uncertain Attribution

Sometimes, the attackers were helped in carrying out the attack by internal participants. In such cases, biometrics will help identify them by their behavioral profile and identify those involved within the organization.

13.12.9 User/Client Carelessness

The human factor assumes that malicious intent is not always necessary for unwanted access. Sometimes, it is enough to be distracted and forget to log out. If the workstation is attacked in this case, behavioral biometrics will quickly take the necessary action if an unauthorized user is detected.

13.12.10 Identity Fraud

In the event of theft of user credentials, organizations that work with the end-user need to suppress attempts to access data or services by criminals. A biometric behavioral profile will also help with this.

13.12.11 License Mismanagement

Sharing of licenses becomes a threat when used illegally. This is a common practice, so behavioral biometrics should be used to eliminate the associated risks. It will ensure that only named persons are using licensed services or products.

13.13 Merits and Demerits

Voice recognition:

Merits: This is one of the most robust and reliable login methods available today. Many technologies have deployed voice recognition, such as online banking, gaming systems, phones, televisions, systems, etc. [1]. Furthermore, voice identification has been implemented in many security sectors, forensic domain, surveillance, etc.

Demerits: Sometimes, the voice recognition system can be affected by a number of real-time factors; throat infections, illness, being under the control of emotions, and issues of aging will impact the accuracy rate.

Gait recognition:

Merits: This is one of the most preferred authentication methods, especially in crowded areas that require a reliable access control system to sensitive buildings; comparing biometrics traits like face, finger and voice are their special features. For example, the effectiveness of individuals' recognition at long distances and less cooperation between the technology itself and the observed object or even no need to have a person's permission [2]. Thus, gait recognition is manifested as a non-intrusive authentication technology. Additionally, it does not require any deployment of a sensor or hardware. Old-style CCTV may work flawlessly. Likewise, the most attractive feature here is the ability to identify a person even with less resolution of the captured images from the selected videos.

Demerits: Although gait recognition has the ability to overcome some difficulties of other biometric traits, there are factors that affect the overall system performance. One of these factors is that an injured leg affects gait recognition performance [2]. In addition, when a person carries an object, wears different types of clothes and views change.

Signature recognition:

Merits: The handwritten signature is one of the most popular ways of authenticating a personal identity. It has a special characteristic among other soft biometrics methodologies in that very low measurement is required while verifying the individual signature; the widespread use of this technology has made it the most familiar and close to everyone's daily life [4]. An attractive feature of signature recognition is the difficulty of taking a person's handwritten signature when they are unconscious. This is unlike other biometrics technologies, such as fingerprint methodology, in which the person's print can be easily collected even when the person is unconscious. The genuine nature of an individual handwritten signature is also an important factor to study this type of biometrics attribute.

Demerits: Here, an individual may change his sketch of the signature over time. Additionally, the state of an individual's health can affect the way of providing their signature.

Keystroke dynamics:

Merits: Here, external hardware components are needed to connect to a system. The only device to be used for data collection is the keyboard. This technology provides an effective extension to the current old password verification mechanism. Furthermore, recognizing individuals based on their typing rhythm is an embedded security technique that makes it difficult to be observed by outsiders; the analysis of keystroke dynamics does not produce a huge computing process.

Demerits: Since it deals with keyboard typing tasks, it requires good typing skills in order to acquire a good feature for each individual.

13.14 Future of Behavioral Biometrics

As artificial intelligence and ML continue to advance, their ability to draw conclusions, learn from, and take intelligent action based on real-time data inputs will also grow. Several business and security applications rely on the automatic recognition of individuals nowadays. Behavioral biometrics identifies/verifies an individual based on the analysis of their distinct physiological or behavioral traits. In general, there is no 100 percent accurate and reliable biometric system to be deployed in the security sectors. Each method has its own merits and demerits. A detailed discussion has been conducted in this work to introduce the most used behavioral biometrics.

References

1. Tandel NH, Prajapati HB, and Dabhi VK. Voice Recognition and Voice Comparison using Machine Learning Techniques: A Survey. 2020 6th International Conference on Advanced Computing and Communication Systems (ICACCS). IEEE; 2020 Mar.
2. Babaee M, Li L, and Rigoll G. Gait Recognition from Incomplete Gait Cycle. 2018 25th IEEE International Conference on Image Processing (ICIP). IEEE; 2018 Oct.
3. Mitchell C, and Shing C-C. Discussing Alternative Login Methods and Their Advantages and Disadvantages. 2018 14th International Conference on Natural Computation, Fuzzy Systems and Knowledge Discovery (ICNC-FSKD). IEEE; 2018 Jul.

4. Bilal Hassan, Ebroul Izquierdo, and Tomas Piatrik. Soft biometrics: a survey Role of Computer Vision in Smart Cities: Applications and Research Challenges Published: 02 March 2021.
5. Sixian Sun, Xiao Fu, Hao Ruan, Xiaojiang Du, Bin Luo, and Mohsen Guizani, Real-Time Behavior Analysis and Identification for Android Application IEEE; VOLUME 6, 2018

Index

Page numbers in *italics* refer figures and **bold** refer tables.

Printed in the United States
by Baker & Taylor Publisher Services